AF595041

Dental Implant Macrogeometry and Biomaterials

Dental Implant Macrogeometry and Biomaterials

Special Issue Editor

José Luis Calvo Guirado

MDPI • Basel • Beijing • Wuhan • Barcelona • Belgrade • Manchester • Tokyo • Cluj • Tianjin

Special Issue Editor
José Luis Calvo Guirado
Universidad Católica San Antonio de Murcia (UCAM)
Spain

Editorial Office
MDPI
St. Alban-Anlage 66
4052 Basel, Switzerland

This is a reprint of articles from the Special Issue published online in the open access journal *Symmetry* (ISSN 2073-8994) (available at: https://www.mdpi.com/journal/symmetry/special_issues/Dental_Implant_Macrogeometry_Biomaterials).

For citation purposes, cite each article independently as indicated on the article page online and as indicated below:

LastName, A.A.; LastName, B.B.; LastName, C.C. Article Title. *Journal Name* **Year**, *Article Number*, Page Range.

ISBN 978-3-03936-453-4 (Pbk)
ISBN 978-3-03936-454-1 (PDF)

Contents

About the Special Issue Editor vii

Preface to "Dental Implant Macrogeometry and Biomaterials" ix

Luca Comuzzi, Margherita Tumedei, Adriano Piattelli and Giovanna Iezzi
Short vs. Standard Length Cone Morse Connection Implants: An In Vitro Pilot Study in Low Density Polyurethane Foam
Reprinted from: *Symmetry* **2019**, *11*, 1349, doi:10.3390/sym11111349 1

Sergio Alexandre Gehrke, Mauro Bercianos, Jorge Gonzalo Aguerrondo, José Luis Calvo-Guirado and Juan Carlos Prados-Frutos
Influence of Mucosal Thickness, Implant Dimensions and Stability in Cone Morse Implant Installed at Subcrestal Bone Level on the Peri-Implant Bone: A Prospective Clinical and Radiographic Study
Reprinted from: *Symmetry* **2019**, *11*, 1138, doi:10.3390/sym11091138 11

Andrés Parrilla-Almansa, Carlos Alberto González-Bermúdez, Silvia Sánchez-Sánchez, Luis Meseguer-Olmo, Carlos Manuel Martínez-Cáceres, Francisco Martínez-Martínez, José Luis Calvo-Guirado, Juan José Piñero de Armas, Juan Manuel Aragoneses, Nuria García-Carrillo and Piedad N. De Aza
Intraosteal Behavior of Porous Scaffolds: The mCT Raw-Data Analysis as a Tool for Better Understanding
Reprinted from: *Symmetry* **2019**, *11*, 532, doi:10.3390/sym11040532 25

Thaiz Carrera-Arrabal, José Luis Calvo-Guirado, Fabricio Passador-Santos, Carlos Eduardo Sorgi da Costa, Frank Róger Teles Costa, Antonio Carlos Aloise, Marcelo Henrique Napimoga, Juan Manuel Aragoneses and André Antonio Pelegrine
Vertical Bone Construction with Bone Marrow-Derived and Adipose Tissue-Derived Stem Cells
Reprinted from: *Symmetry* **2019**, *11*, 59, doi:10.3390/sym11010059 41

Enrique Fernández-Bodereau, Viviana Yolanda Flores, Rafael Arcesio Delgado-Ruiz, Juan Manuel Aragoneses and José Luis Calvo-Guirado
Evaluation of the Cortical Deformation Induced by Distal Cantilevers Supported by Extra-Short Implants: A Finite Elements Analysis Study
Reprinted from: *Symmetry* **2018**, *10*, 762, doi:10.3390/sym10120762 53

About the Special Issue Editor

José Luis Calvo Guirado is a dual doctor in dentistry and bioengineering of biomaterials. He also is Full Professor and Chairman of Oral Surgery and Implant Dentistry at San Antonio Catholic University of Murcia (UCAM) in Spain and Full Professor in the General and Implant Dentistry Master Course at Murcia University in Spain since 2001 to 2015. Since 2015, he has been its director. He is the Director of International Dentistry Research Cathedra and Director of the research Group at UCAM University. He is a Research Professor at the School of Dental Medicine at Stony Brook University in New York, Visiting Professor of Oral Surgery and Implant Dentistry at Belgrade University in Serbia, Visiting professor of Implant Dentistry Specialization at Catholic University of Córdoba Argentina, and Visiting Professor of Tito Maiorescu University of Bucharest, Rumania. His recognized international expertise is based in oral surgery, prosthodontics, dental implants, bone substitutes, animal research, and digital technologies. He is author of more than 234 publications in PubMed and 350 manuscripts in non-JCR Journals. He is an author of 7 books and a co-author of another 8 scientific books. He is Vice President of the Implant Research Group of International Association of Dental Research (IADR). He is ranked the 41st worldwide in dentistry research in the ExpertScape Web Ranking.

Preface to "Dental Implant Macrogeometry and Biomaterials"

This book is the result of the publication of a Special Issue published in the MDPI journal Symmetry, entitled "Dental Implant Macrogeometry and Biomaterials" (https://www.mdpi.com/journal/symmetry/special). We are honored by the interest taken in this Special Issue by academics, researchers, and specialists. It was finalized with a total of 5 published articles. This book is released due to the efforts of the Dentistry International Cooperation between scientists from Argentina, Italy, Brazil, Spain, Uruguay, and the Dominican Republic. The topics covered in the aforementioned review or research articles comprise different interest areas in the field of dental implant design, biomaterial synthesis, processing, and complex characterization. We have also included some digital papers in order to cover the entry of hole spectrometry into the medical field.

José Luis Calvo Guirado
Special Issue Editor

Article

Short vs. Standard Length Cone Morse Connection Implants: An In Vitro Pilot Study in Low Density Polyurethane Foam

Luca Comuzzi [1,†], Margherita Tumedei [2,*,†], Adriano Piattelli [2,3,4] and Giovanna Iezzi [2]

1 Private practice, via Raffaello 36/a, 31020 San Vendemiano (TV), Italy; luca.comuzzi@gmail.com
2 Department of Medical, Oral and Biotechnological Sciences, University "G. D'Annunzio" of Chieti-Pescara, 66100 Chieti, Italy; apiattelli@unich.it (A.P.); g.iezzi@unich.it (G.I.)
3 Catholic University of San Antonio de Murcia (UCAM), Av. de los Jerónimos, Guadalupe, 135 30107 Murcia, Spain
4 Villaserena Foundation for Research, 65121 Città Sant'Angelo (Pescara), Italy
* Correspondence: margytumedei@yahoo.it; Tel.: +39-0871-3554083
† These two Authors had an equal contribution to the study. LC for the execution of the experimental portion, MT for data collection, data analysis, statistical evaluation, final editing of the manuscript.

Received: 9 October 2019; Accepted: 30 October 2019; Published: 1 November 2019

Abstract: The aim of the investigation was to evaluate the insertion torque, pull-out torque and implant stability quotient (ISQ) of short implants (SI) and standard length implants (ST) inserted into linearly elastic and constitutive isotropic symmetry polyurethane foam blocks. Short dental titanium implants with a Cone Morse connection and a conical shape (test implants: Test Implant A—diameter 5.5 mm and length 6 mm) (Test Implant B—diameter 5.5 mm and length 5 mm) were used for the present in vitro investigation. ST implants (4 mm diameter and 10 mm length), with a Cone Morse connection and a conical shape, were used as Control Implant A and as Control Implants B. These two latter implants had a different macro design. A total of 20 implants (5 Test A, 5 Test B, 5 Control A and 5 Control B) were used for the present research. The results were similar when comparing the Test A and Test B implants. The test implants had very good stability in polyurethane 14.88–29.76 kgm^3 density blocks. The insertion torque values were very high for both types of test implant (25–32 Ncm on 14.88 kgm blocks, and up to 45 Ncm in 29.76 kgm^3 blocks). The pull-out test values were very similar to the insertion torque values. The ISQ values were significantly high with 75–80 in 14.88 kgm^3 blocks, and 78–83 in 29.76 kgm^3 blocks. No differences were found in the values of the Control A and Control B implants. In both these implants, the insertion torque was quite low in the 14.88 kgm^3 blocks (16–28 Ncm). Better results were found in the 29.76 kgm^3 blocks. The pull-out values for these control implants were slightly lower than the insertion torque values. High ISQ values were found in both control implants (57–80). When comparing SI and ST implants, the SI had a similar if not better performance in low quality polyurethane foam blocks (14.88–29.76 kgm), corresponding to D3 and D4 bone.

Keywords: bone density; implant stability quotient; insertion torque; polyurethane foam blocks; pull-out torque; resonance frequency analysis; short implants; standard length implants

---

1. Introduction

Short implants (SI), defined in recent years as implants of less than 10 mm in length [1], seem to have some advantages in certain clinical situations, such as atrophy of the alveolar processes, poor bone quality, and pneumatization of the maxillary sinus [2]. SI are less invasive, simpler to use in the hands of the average clinician, and their surgery is shorter, with a lower morbidity, lower costs and lesser biological complications [1,3–7]. More recently, Ultrashort or Extra-short (<6 mm) implants have been proposed [8–11]. Alternatives to the use of SI are sinus augmentation procedures, the use of zygomatic implants, guided bone regeneration procedures, onlay grafts, inlay grafts, distraction osteogenesis, and lateralization of the inferior alveolar nerve [4,12]. Several recent systematic reviews, some of them with a metanalysis of the data, have shown:

no differences in the in the survival rate between SI and standard length implants (ST);
no differences in marginal bone loss (MBL);
lower biological complications in SI;
good primary stability in SI;
higher mechanical complications in SI [3,4,9,10,13–16].

Primary dental implant stability (PS), i.e., an absence of micromotion of the implant immediately after implant placement, has been reported to have an important role in implant osseointegration [17–19]. PS seemed to be closely correlated to bone quality and quantity, implant macrostructure, implant length and diameter, surgical technique, and the fitting of the implant into the site [17–21]. Bone density has been correlated to the amount of bone-to-implant contact (BIC) [19], and BIC to the PS [17]. Bone density has been measured with the use of different techniques: insertion torque (IT), removal torque (RT), and resonance frequency analysis (RFA), producing a value giving the implant stability quotient (ISQ) [22]. Polyurethane foam has been recognized as a standard material for testing instruments by the American Society for Testing and Materials (ASTM F-1839-08) ("Standard specification for Rigid Polyurethane Foam for Use as a Standard Material for Test Orthopaedic Devices for Instruments"). Polyurethane foam has been widely used as an alternative material in biomechanical tests evaluating, for example, dental implants. It presents consistent mechanical characteristics, has features similar to bone tissue, is very reliable and easy to use, requiring no special handling, and is characterized by linearly elastic and constitutive isotropic symmetry [22–24].

The scope of the present pilot study was to evaluate the insertion torque, pull-out torque and ISQ of SI and ST implants, positioned into polyurethane foam blocks.

2. Materials and Methods

2.1. Dental Implants

The short dental titanium implants with a Cone Morse connection and a conical shape (Implacil De Bortoli, Sao Paulo, Brasil) (Test Implants: Test Implant A—diameter 5.5 mm and length 6 mm) (Test Implant B—diameter 5.5 mm and length 5 mm) were used for the in vitro experimental study. Universal II (UN II) implants (4 mm diameter and 10 mm length), with a Cone Morse connection and a conical shape (Implacil De Bortoli, Sao Paulo, Brasil), were used as Control Implant A, and Universal III (UN III) implants, also with a Cone Morse connection and a conical shape (Implacil De Bortoli, Sao Paulo, Brasil), were used as Control Implants B. These two latter implants differed in their macro design. The UN III macro design differed from the UN II implants regarding its larger thread, the lack of double thread pitch, having a round apex not self-tapping, and in the chambers' patterns between the cutting surface of the threads.

2.2. Study Design

A total of 20 implants (5 Test A, 5 Test B, 5 Control A and 5 Control B) were used in the present investigation. The control implants were inserted following the protocol of the manufacturer: implant lance drill, 2 mm drill (1200 rpm) and 3.5 mm final drill (800 rpm) (Figures 1 and 2). The test implants were inserted following the protocol of the manufacturer: implant lance drill, 2 mm drill (1200 rpm), 3.5 mm conical drill for SI and 4.5 mm final conical drill for SI (800 rpm).

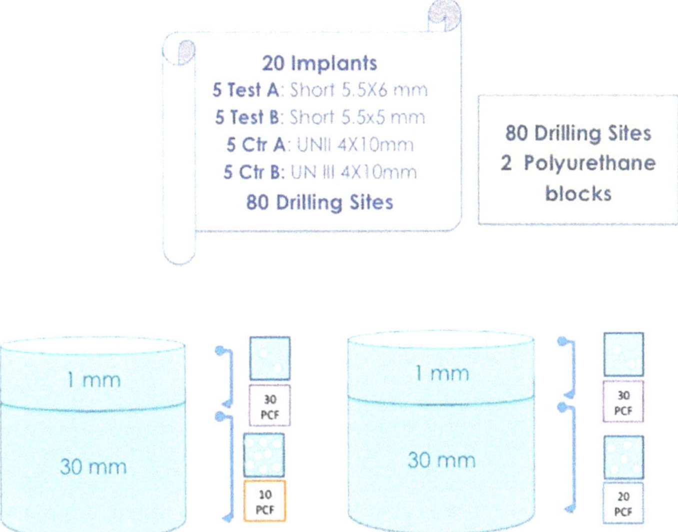

Figure 1. Study design and the experimental groups of the present in vitro investigation. PCF = pound per cubic foot.

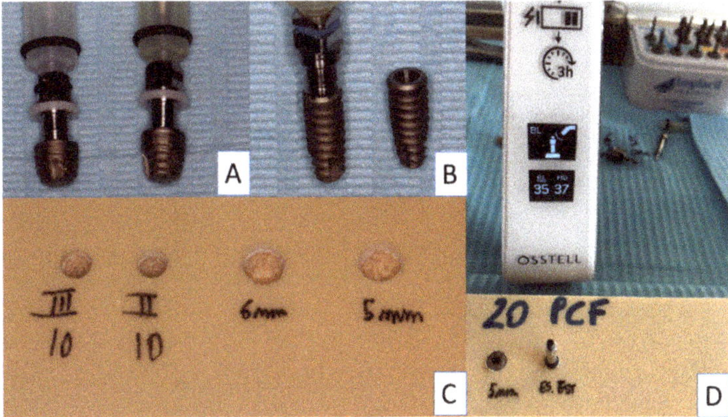

Figure 2. Sequence of the study investigation: (**A**) Test A and Test B; (**B**) Control A and Control B; (**C**) Polyurethane foam block after drilling; (**D**) implant stability quotient (ISQ) measurement device.

The insertion of the implants was made by the handpiece calibrated to 20 rpm speed and a torque of 30 Ncm. The torque peaks were recorded by the software package (ImpDat Plus, East Lansing, Michigan) installed on a digital card. The insertion torque (IT, Ncm) peaks indicated the force of the maximum clockwise movement of the dental fixture positioned into the material. The research was performed by a single operator (LC), recording the fixture insertion and the pull-out torque peaks of the Test A, Test B, Control A and Control B Implants positioned into polyurethane foam blocks in different sizes and densities.

Different types of solid rigid polyurethane foam (SawBones H, Pacific Research Laboratories Inc, Vashon, Washington, USA) with homogeneous densities were selected for the present investigation. The polyurethane foam blocks presented a size of "120 mm × 70 mm × 31 mm".

The block densities of polyurethane samples used in the present investigation were: 16.01 kgm^3 (10 PCF), similar to D3 bone quality, 32.02 kgm^3 (20 PCF), corresponding with and similar to D2 bone;

moreover, a 1 mm sheet of polyurethane with a 48.03 kgm^3 (30 PCF) density, similar to D1 bone, was present to simulate a layer of cortical bone. Ten implant site perforations were performed for each type of implant (Test A, Test B, Control A and Control B) for both polyurethane densities (14.88–29.76 kgm^3), for a total of 80 implant site preparations.

2.3. *Implant Drill*

Test A and Test B implants were inserted following a suggested drill protocol using a lance drill, then a 2 mm bur at 1200 rpm, then a 3.5 mm conical bur, and subsequently a 4.5 mm conical bur (both at 300 rpm) with the implant insertion at 20 Rpm. Control A and Control B Implants were inserted using a surgical lance drill, then a 2 mm bur, and subsequently a conical 3.5 mm bur at 800 Rpm with the implant insertion at 20 rpm.

2.4. *Insertion Torque and Pull-Out Torque*

The comparative research evaluating the insertion torque and pull-out peaks was conducted using a calibrated torque ratchet (Implacil De Bortoli, Sao Paulo, Brasil) provided by a torque range of 5–80 N/cm. The final 1 mm insertion torque of the implants into the polyurethane blocks was evaluated using a calibrated torque ratchet (Implacil De Bortoli, Sao Paulo, Brasil). In the present investigation, the mechanical torque gauges were used to assess the insertion torque and the pull-out strength values.

2.5. *Resonance Frequency Evaluation*

After the fixture positioning, the primary stability was evaluated using Resonance Frequency analysis (RFA) values expressed in the implant stability quotient (ISQ) by a hand-screwed Smart-Pegs type 7 for test implants (Osstell Mentor Device, Integration Diagnostic AB, Savadelen, Sweden) (Figure 2). The ISQ values ranged from 0 to 100 (measured by a frequency in the range 3500–85,000 Hz), and was classified into Low (less than 60 ISQ), Medium (in the range 60–70 ISQ), and High stability rate (more than 70 ISQ) [25]. Moreover, RFA evaluation was repeated twice for each sample evaluated. The RFA evaluation was performed following two different orientations separated by a 90-degree angle, and the mean ISQ peaks were calculated.

2.6. *Statistical Analysis*

The Shapiro–Wilks test was performed to evaluate data normality. Moreover, the differences between the peaks of insertion torque, pull-out strength and the RFA of the study groups were evaluated using a two-way analysis of variance (ANOVA), followed by the Tukey post-hoc test. A p-value < 0.05 was considered statistically significant. The research data and the statistical analysis were performed using the software package Excel (Microsoft Office, Redmond, USA) and GraphPad 6 (Prism, San Diego, USA).

3. Results

The results are similar when comparing the Test A and Test B implants. The test implants had very good stability in polyurethane 14.88–29.76 kgm^3 density blocks (Figures 3 and 4).

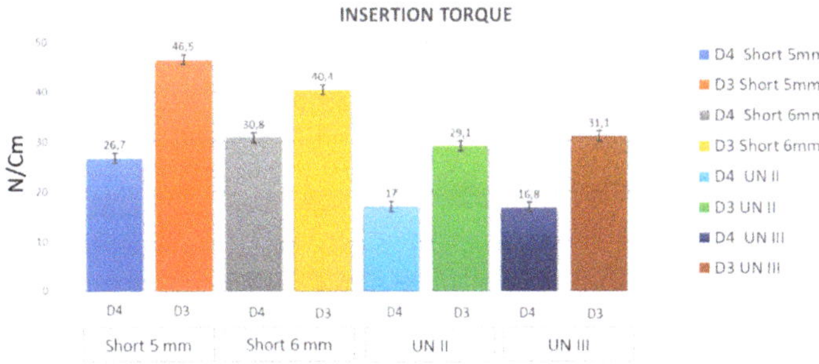

Figure 3. Insertion torque evaluation of the study groups: Test A, Test B. Control A and Control B.

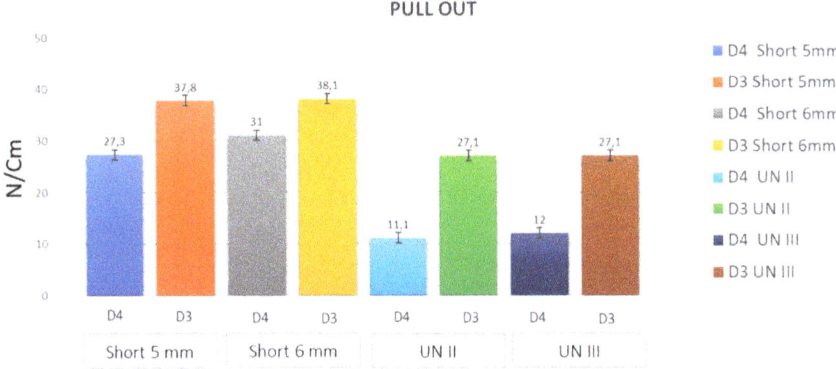

Figure 4. Pull-out outcome of the implant positioned into the polyurethane foam block: Test A, Test B. Control A and Control B.

The insertion torque values were very high for both types of test implant (25–32 Ncm on 14.88 kgm^3 blocks, and up to 45 Ncm in 29.76 kgm^3 blocks) (Figure 3; Tables 1 and 2).

The pull-out test values were very similar to the insertion torque values (Figure 4). The ISQ values were significantly high with 75–80 in 14.88 kgm^3 blocks, and 78–83 in 29.76 kgm^3 blocks (Tables 3 and 4).

Table 1. Summary of the insertion torque value for the implant positioned into polyurethane foam (14.88–29.76 kgm^3); Test A, Test B. Control A and Control B.

Insertion Torque	D4 Density				D3 Density			
	Short 5 mm (A)	Short 6 mm (B)	UN II (C)	UN III (D)	Short 5 mm (E)	Short 6 mm (F)	UN II (G)	UN III (H)
Mean	26.7	30.8	17	16.8	46.5	40.4	29.1	31.1
Std. Deviation	±1.059	±1.033	±0.942	±1.135	±1.269	±1.350	±0.994	±0.994

Table 2. Insertion torque: ANOVA Bonferroni post-hoc comparison test (CI = Confidence Interval; Diff = Difference).

Multiple Comparison Insertion Torque	95.00% CI of Diff,	Adjusted p Value
A-B	−5.577 to −2.623	<0.0001
B-C	12.32 to 15.28	<0.0001
C-D	−1.277 to 1.677	N.S.D.
D-E	−31.18 to −28.22	<0.0001
A-D	8.423 to 11.38	<0.0001
A-C	8.223 to 11.18	<0.0001
B-D	12.52 to 15.48	<0.0001
E-F	4.623 to 7.577	<0.0001
F-G	9.823 to 12.78	<0.0001
G-H	−3.477 to −0.5230	0.0006
E-H	13.92 to 16.88	<0.0001
F-H	7.823 to 10.78	<0.0001
E-G	15.92 to 18.88	<0.0001

N.S.D. (no significant differences).

Table 3. Summary of the pull-out outcome of the implant positioned into polyurethane foam (14.88 D3 Density–29.76 kgm^3 D3 Density); Test A, Test B. Control A and Control B.

Pull Out	D4 Density				D3 Density			
	Short 5 mm (A)	Short 6 mm (B)	UN II (C)	UN III (D)	Short 5 mm (E)	Short 6 mm (F)	UN II (G)	UN III (H)
Mean	27.3	31	11.1	12	37.8	38.1	27.1	27.1
Std. Deviation	±1.059	±1.054	±0.994	±1.247	±0.918	±0.875	±1.101	±1.287

Table 4. Pull-out: ANOVA Bonferroni post-hoc comparison test (CI = Confidence Interval; Diff = Difference).

Multiple Comparison Pull Out	95.00% CI of Diff,	Adjusted p Value
A-B	−5.124 to −2.276	<0.0001
B-C	18.48 to 21.32	<0.0001
C-D	−2.324 to 0.5237	N.S.D.
A-D	13.88 to 16.72	<0.0001
B-D	17.58 to 20.42	<0.0001
A-C	14.78 to 17.62	<0.0001
D-E	−27.24 to −24.36	<0.0001
E-F	−1.724 to 1.124	>0.9999
F-G	9.576 to 12.42	<0.0001
G-H	−1.424 to 1.424	N.S.D.
E-H	9.276 to 12.12	<0.0001
F-H	9.576 to 12.42	<0.0001
E-G	9.276 to 12.12	<0.0001
A-B	−5.124 to −2.276	<0.0001

N.S.D. (no significant differences).

The clinical sensation of the very good stability of the implant was felt at the initial insertion into the polyurethane blocks. No differences were found in the values of the Control A and Control B implants (Tables 1–4). In both these implants, the insertion torque was quite low in the 14.88 kgm^3 blocks (16–28 Ncm) (Tables 1 and 2). Better results were found in the 29.76 kgm^3 blocks, with significantly good stability of the implants. The pull-out values for these control implants were slightly lower than the insertion torque values. High ISQ values were found in both control implants (57–80) (Figure 5; Tables 5 and 6).

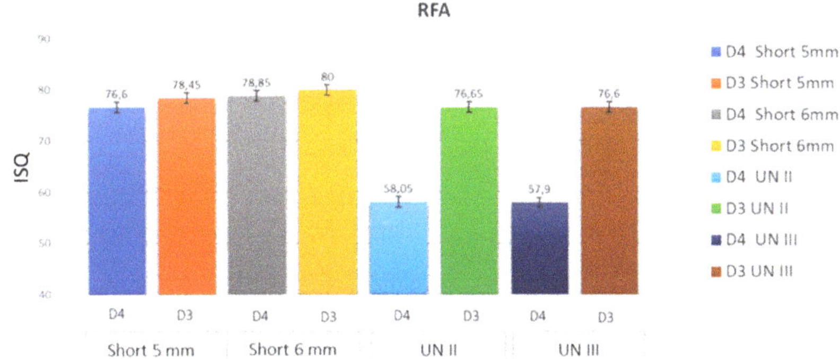

Figure 5. ISQ outcome of the implant positioned into polyurethane foam block: Test A, Test B. Control A and Control B.

Table 5. Summary of the ISQ of the implant positioned into Polyurethane foam (14.88 D4 Density–29.76 kgm^3 D3 Density); Test A, Test B. Control A and Control B.

ISQ	D4 Density				D3 Density			
	Short 5 mm (A)	Short 6 mm (B)	UN II (C)	UN III (D)	Short 5 mm (E)	Short 6 mm (F)	UN II (G)	UN III (H)
Mean	76.6	78.85	58.05	57.9	78.45	80	76.65	76.6
Std. Deviation	±0.966	±0.747	±0.283	±0.809	±0.550	±1.491	±1.001	±0.774

Table 6. ISQ: ANOVA Bonferroni Post-hoc comparison test (CI = Confidence Interval; Diff = Difference).

Multiple Comparison RFA	95.00% CI of Diff,	Adjusted *p* Value
A-B	−3.430 to −1.070	<0.0001
B-C	19.62 to 21.98	<0.0001
C-D	−1.030 to 1.330	N.S.D.
A-D	17.52 to 19.88	<0.0001
B-D	19.77 to 22.13	<0.0001
A-C	17.37 to 19.73	<0.0001
D-E	−21.74 to −19.36	<0.0001
E-F	−2.730 to −0.3703	0.0027
F-G	2.170 to 4.530	<0.0001
G-H	−1.130 to 1.230	N.S.D.
E-H	0.6703 to 3.030	0.0002
F-H	2.220 to 4.580	<0.0001
E-G	0.6203 to 2.980	0.0003

N.S.D. (no significant differences).

4. Discussion

The atrophy of the posterior regions of the jaws with reduced bone quality and quantity could limit the use of standard length implants (≥10 mm), without doing an invasive sinus grafting procedure. Recently, it has been reported in a few systematic reviews with meta-analysis that, in these cases, short implants could be a suitable alternative [2–8,12,13]. These reviews have reported, for short implants, survival rates similar to those of standard length implants and the capability to osseointegrate and to bear a functional load [8]. In recent years, a reduction in the implant length of short implants has been reported in the literature [1,7,8,12]. Polyurethane foam could be an alternative useful material to provide biomechanical tests substituting, for example, animal bone. ("Standard specification for Rigid Polyurethane Foam for Use as a Standard Material for Test Orthopaedic Devices for Instruments"). Polyurethane presents a cellular structure with constant mechanical characteristics, and similar properties to bone. In the present study, very good stability was obtained for both test and control implants. In the test implants, insertion torque, pull-out torque and ISQ values were all very high, showing the very good stability of both types of SI. Also, all the values for the ST implants were quite high, with better results, and then better stability, in higher density polyurethane blocks. The density of polyurethane blocks is similar to the structure of the bone in the posterior regions in humans. The conical shape of all these implants was probably instrumental in achieving such good levels of stability.

The main reason for measuring the implant primary stability concerns the ability to predict the prognosis of the dental implant procedure. Comuzzi et al., in vitro, reported that in polyurethane foam, ISQ, insertion torque, and pull-out measuring provide the high repeatability and reproducibility that represent suitable indicators for implant stability [26]. In the present investigation, the implant primary stability was evaluated in a controlled reproducible study design and without the variables correlated to the use of animal bone.

The study effectiveness showed that the types of polyurethane used in the present in vitro study were shown to be constituted by a homogenous material.

Moreover, implants with a conical shape have a high stability even in blocks with a low density (D4 Density) and no differences were found between 5 mm and 6 mm long implants, where reasonable values of insertion torque and pull-out tests were found in short implants in both polyurethane densities. Comuzzi et al. reported in vitro that a conical shape, rather than a cylindrical design, provided increased values of insertion torque and pull-out strength [26]. Thus, the ISQ values of both short implants were great, with an increased clinical sensation of the high stability of the implants.

The posterior maxilla is an anatomical region often characterized by poor bone quality and quantity. Thus, the 5 mm short implant could probably be more useful for insertion in this area in cases of reduced bone volume and density.

In fact, the choice of a short implant is clinically indicated as an alternative to more invasive regenerative procedures in cases of bone atrophy of the posterior ridge's regions [3].

The standard length implants had slightly lower values of insertion torque and pull-out torque, and very high ISQ values, where the conometric connection of all the types of implants seemed to also resist quite well to very high torque values (up to 60–80 Ncm).

5. Conclusions

In conclusion, when comparing the SI and ST, the SI had a similar if not better performance in low quality polyurethane foam blocks corresponding to D3 and D4 bone.

Author Contributions: A.P. Conceptualization, L.C. Investigation, A.P., L.C. Methodology; Supervision G.I., Validation A.P., G.I.; M.T. Data Curation, M.T. Formal Analysis; and A.P. and M.T. Writing Original Draft, A.P. and M.T. Writing Review Draft.

Funding: This study was supported by the Italian Ministry of Education, University and Research (M.I.U.R.), Rome, Italy, and by a grant from Implacil De Bortoli, Sao Paulo, Brasil.

Conflicts of Interest: The authors declare no conflict of interest.

References

1. Lombardo, G.; Pighi, J.; Marincola, M.; Corrocher, G.; Simancas-Pallares, M.; Nocini, P.F. Cumulative Success Rate of Short and Ultrashort Implants Supporting Single Crowns in the Posterior Maxilla: A 3-Year Retrospective Study. *Int. J. Dent.* **2017**, *2017*, 8434281. [CrossRef] [PubMed]
2. Nielsen, H.B.; Schou, S.; Isidor, F.; Christensen, A.E.; Starch-Jensen, T. Short implants (≤8 mm) compared to standard length implants (>8 mm) in conjunction with maxillary sinus floor augmentation: A systematic review and meta-analysis. *Int. J. Oral Maxillofac. Surg.* **2019**, *48*, 239–249. [CrossRef] [PubMed]
3. Cruz, R.S.; Lemos, C.A.A.; Batista, V.E.S.; Oliveira, H.F.F.E.; Gomes, J.M.L.; Pellizzer, E.P.; Verri, F.R. Short implants versus longer implants with maxillary sinus lift. A systematic review and meta-analysis. *Braz. Oral Res.* **2018**, *32*, 86. [CrossRef] [PubMed]
4. Tolentino da Rosa de Souza, P.; Binhame Albini Martini, M.; Reis Azevedo-Alanis, L. Do short implants have similar survival rates compared to standard implants in posterior single crown?: A systematic review and meta-analysis. *Clin. Implant. Dent. Relat. Res.* **2018**, *20*, 890–901. [CrossRef] [PubMed]
5. Lemos, C.A.A.; Ferro-Alves, M.L.; Okamoto, R.; Mendonça, M.R.; Pellizzer, E.P. Short dental implants versus standard dental implants placed in the posterior jaws: A systematic review and meta-analysis. *J. Dent.* **2016**, *47*, 8–17. [CrossRef] [PubMed]
6. N Dias, F.J.; Pecorari, V.G.A.; Martins, C.B.; Del Fabbro, M.; Casati, M.Z. Short implants versus bone augmentation in combination with standard-length implants in posterior atrophic partially edentulous mandibles: Systematic review and meta-analysis with the Bayesian approach. *Int. J. Oral Maxillofac. Surg.* **2019**, *48*, 90–96. [CrossRef]
7. Markose, J.; Eshwar, S.; Srinivas, S.; Jain, V. Clinical outcomes of ultrashort sloping shoulder implant design: A survival analysis. *Clin. Implant Dent. Relat. Res.* **2018**, *20*, 646–652. [CrossRef]
8. Urdaneta, R.A.; Daher, S.; Leary, J.; Emanuel, K.M.; Chuang, S.K. The survival of ultrashort locking-taper implants. *Int. J. Oral Maxillofac. Implant* **2012**, *27*, 644–654.
9. Ravidà, A.; Barootchi, S.; Askar, H.; Suárez-López Del Amo, F.; Tavelli, L.; Wang, H.L. Long-Term Effectiveness of Extra-Short (≤6 mm) Dental Implants: A Systematic Review. *Int. J. Oral Maxillofac. Implant.* **2019**, *34*, 68–84. [CrossRef]
10. Bitaraf, T.; Keshtkar, A.; Rokn, A.R.; Monzavi, A.; Geramy, A.; Hashemi, K. Comparing short dental implant and standard dental implant in terms of marginal bone level changes: A systematic review and meta-analysis of randomized controlled trials. *Clin. Implant Dent. Relat. Res.* **2019**, *21*, 796–812. [CrossRef]
11. Deporter, D. *Short and Ultrashort Implants*; Quintessence Publishing: New Malden, UK, 2018; pp. 59–74.
12. Deporter, D.; Ogiso, B.; Sohn, D.S.; Ruljancich, K.; Pharoah, M. Ultrashort sintered porous-surfaced dental implants used to replace posterior teeth. *J. Periodontol.* **2008**, *79*, 1280–1286. [CrossRef] [PubMed]
13. Fan, T.; Li, Y.; Deng, W.W.; Wu, T.; Zhang, W. Short Implants (5 to 8 mm) Versus Longer Implants (>8 mm) with Sinus Lifting in Atrophic Posterior Maxilla: A Meta-Analysis of RCTs. *Clin. Implant Dent. Relat. Res.* **2017**, *19*, 207–215. [CrossRef] [PubMed]
14. Al-Johany, S.S. Survival Rates of Short Dental Implants (≤6.5 mm) Placed in Posterior Edentulous Ridges and Factors Affecting their Survival after a 12-Month Follow-up Period: A Systematic Review. *Int. J. Oral Maxillofac. Implant* **2019**, *34*, 605–621. [CrossRef] [PubMed]
15. Martinolli, M.; Bortolini, S.; Natali, A.; Pereira, L.J.; Castelo, P.M.; Rodrigues Garcia, R.C.M.; Gonçalves, T.M.S.V. Long-term survival analysis of standard-length and short implants with multifunctional abutments. *J. Oral Rehabil.* **2019**, *46*, 640–646. [CrossRef]
16. Felice, P.; Soardi, E.; Pellegrino, G.; Pistilli, R.; Marchetti, C.; Gessaroli, M.; Esposito, M. Treatment of the atrophic edentulous maxilla: Short implants versus bone augmentation for placing longer implants. Five-month post-loading results of a pilot randomised controlled trial. *Eur. J. Oral Implant* **2011**, *4*, 191–202.
17. Gehrke, S.A.; Guirado, J.L.C.; Bettach, R.; Fabbro, M.D.; Martínez, C.P.A.; Shibli, J.A. Evaluation of the insertion torque, implant stability quotient and drilled hole quality for different drill design: An in vitro Investigation. *Clin. Oral Implant Res.* **2018**, *29*, 656–662. [CrossRef]

18. Romanos, G.E.; Delgado-Ruiz, R.A.; Sacks, D.; Calvo-Guirado, J.L. Influence of the implant diameter and bone quality on the primary stability of porous tantalum trabecular metal dental implants: An in vitro biomechanical study. *Clin. Oral Implant Res.* **2018**, *29*, 649–655. [CrossRef]
19. Möhlhenrich, S.C.; Heussen, N.; Elvers, D.; Steiner, T.; Hölzle, F.; Modabber, A. Compensating for poor primary implant stability in different bone densities by varying implant geometry: A laboratory study. *Int. J. Oral. Maxillofac. Surg.* **2015**, *44*, 1514–1520. [CrossRef]
20. Yamaguchi, Y.; Shiota, M.; FuJii, M.; Sekiya, M.; Ozeki, M. Development and application of a direct method to observe the implant/bone interface using simulated bone. *Springerplus* **2016**, *5*, 494. [CrossRef]
21. Falco, A.; Berardini, M.; Trisi, P. Correlation Between Implant Geometry, Implant Surface, Insertion Torque, and Primary Stability: In Vitro Biomechanical Analysis. *Int. J. Oral Maxillofac. Implant* **2018**, *33*, 824–830. [CrossRef]
22. Di Stefano, D.A.; Arosio, P.; Gastaldi, G.; Gherlone, E. The insertion torque-depth curve integral as a measure of implant primary stability: An in vitro study on polyurethane foam blocks. *J. Prosthet. Dent.* **2018**, *120*, 706–714. [CrossRef] [PubMed]
23. Tsolaki, I.N.; Tonsekar, P.P.; Najafi, B.; Drew, H.J.; Sullivan, A.J.; Petrov, S.D. Comparison of Osteotome and Conventional Drilling Techniques for Primary Implant Stability: An In Vitro Study. *J. Oral Implant* **2016**, *42*, 321–325. [CrossRef] [PubMed]
24. Oliveira, P.S.; Rodrigues, J.A.; Shibli, J.A.; Piattelli, A.; Iezzi, G.; Perrotti, V. Influence of osteoporosis on the osteocyte density of human mandibular bone samples: A controlled histological human study. *Clin. Oral Implant Res.* **2016**, *27*, 325–328. [CrossRef] [PubMed]
25. Sennerby, L.; Meredith, N. Implant stability measurements using resonance frequency analysis: Biological and biomechanical aspects and clinical implications. *Periodontology* **2008**, *47*, 51–66. [CrossRef]
26. Comuzzi, L.; Iezzi, G.; Piattelli, A.; Tumedei, M. An In Vitro Evaluation, on Polyurethane Foam Sheets, of the Insertion Torque (IT) Values, Pull-Out Torque Values, and Resonance Frequency Analysis (RFA) of NanoShort Dental Implants. *Polymer* **2019**, *11*, 1020. [CrossRef]

Article

Influence of Mucosal Thickness, Implant Dimensions and Stability in Cone Morse Implant Installed at Subcrestal Bone Level on the Peri-Implant Bone: A Prospective Clinical and Radiographic Study

Sergio Alexandre Gehrke [1,2,*], Mauro Bercianos [3], Jorge Gonzalo Aguerrondo [3], José Luis Calvo-Guirado [4] and Juan Carlos Prados-Frutos [5]

1 Biotecnos Research Center, 11100 Montevideo, Uruguay
2 Master Postgraduate Program in Biotechnology of UCAM, 11100 Montevideo, Uruguay
3 Postgraduate Program in Biotechnology of UCAM, 11100 Montevideo, Uruguay
4 Oral and Implant Surgery, Faculty of Health Sciences, Universidad Católica San Antonio de Murcia (UCAM), 30107 Murcia, Spain
5 Department of Medicine and Surgery, Rey Juan Carlos University, Alcorcón, 28922 Madrid, Spain
* Correspondence: sergio.gehrke@hotmail.com

Received: 12 July 2019; Accepted: 28 August 2019; Published: 7 September 2019

Abstract: The objective of this observational clinical study was to analyze the behavior of peri-implant tissues around cone Morse dental implants installed in the subcrestal bone position considering different clinical variables: Mucosal thickness, implant diameter, and implant length. Thirty patients were selected and included in the present study. Initially the thickness of the mucosa was measured by periapical radiographic and clinically (after the mucosal displaced). According to the planning for each treatment, implants with different dimensions (in length and diameter) were selected and used. Periapical radiographs were obtained at different times: Immediate postoperative (time t1) and 90 days after implantation (time t2). The initial stability of the implants (ISQ) was measured immediately of the implant insertion and 90 days after. The means and standard deviations of the ISQ values were in time t1 was 63.2 ± 6.99 (95% confidence interval (CI): 41 to 83) and in time t2 was 69.7 ± 7.09 (95% CI: 61 to 87). Overall mean of mesial and distal bone loss 90 days after the implantations were 1.11 ± 1.16 mm and 1.11 ± 1.15 mm, respectively. When the variables were considered, in all situations proposed, the bone loss showed differences statistically significant. In conclusion, the implant diameter and mucosal thickness variables showed an important effect on bone loss values. However, the implant length did not show an effect on the peri-implant behavior.

Keywords: crestal bone; cone Morse implants; mucosal thickness; implant dimensions; resonance frequency analysis

1. Introduction

Implantology as a surgical procedure implies the management of a wound involving the soft and hard tissues. After the implant osseointegration, the bone and the mucosa required by its new function of protection of underlying peri-implant structures, is transformed into the peri-implantar mucosa acquiring particular morphological characteristics. For this new sealing function, the epithelial and connective tissues require an appropriate dimension, and if it does not exist it will be created at the expense of bone resorption. Berglund and Lindhe [1] carried out a study with the purpose of confirming this concept, where thinning the tissues also proves that the conformation of the seal requires a minimal mucosal dimension, otherwise it would be created at the expense of bone resorption. In this way, the biology demands a dimension of minimal epithelial and connective tissue adequate for the protection of the underlying structures. In this sense, other authors begin to give importance to the mucosal thickness as a relevant factor independent of the aforementioned bone resorption, using in its clinical trials different types of technology in the area of the implant connection, noticing the inefficacy of these technologies in the control of crestal bone resorption when the mucosa shows little thickness [2–7].

In addition, other factors may affect the behavior of peri-implant tissues, such as: Microgap, micromovement, microtopography of the interface, repeated removal of the abutment, and the platform design (switching or no) [8–11]. Several studies carried out by our group demonstrated that Morse taper implants present a better condition and behavior when compared to implants of internal and external connection, referring to the factors previously described [12,13]. In this way, a recent important systematic review that was published about the performance of the Morse taper connection [14] showed evidence that this type of connection appeared to be superior in terms of bacterial sealing compared with the traditional ones emphasizing that no connection has a 100% bacterial sealing. Morse taper connection systems appear to be more resistant to abutment movement and increased under load space compared to internal and external hexagon implants [15]. Moreover, the Morse taper connection have greater resistance to torque loss than other connection models [16]. This system seems to have less tension on the abutment screw, the cone compensates for the high stresses and protects the screw from overload [17].

Marginal bone stability around dental implants has always been considered one of the main criteria for defining implant success [18]. Then, with the current advancements and new technologies in implant dentistry, we should strive both for bone loss close to zero and to seek out variables that cause higher or lower rates of resorption. Albrektsson et al. reported that the extensive bone resorption after the first year is generally due to an exacerbation of adverse body reactions caused by non-optimal implant components, adverse surgery or prosthodontics, and/or compromised patient factors [19]. In a recent review study of the evidence regarding marginal bone loss around dental implants, Sasada and Cochran concluded that there is a strong indication that contaminated implant-abutment connections may have an effect on peri-implantitis and failure over time [20].

Although the in vitro results show better results in implants of conical internal connection and in vivo results with lower marginal bone loss, all models show comparable rates in terms of implant success and survival. However, these Morse tapered implants need to be evaluated for their clinical behavior in relation to peri-implant tissues, as many manufacturers recommend their infra-osseous installation without explaining the need for this procedure. In this sense, the aim of the present study was to evaluate the clinical performance of an implant with Morse taper connection (submerged 2 mm infra-osseous) and comparing different clinical variables (mucosal thickness, implant dimensions, and implant stability) with the marginal bone behavior. It was hypothesized that mucosal thickness plays a fundamental role in the maintenance of peri-implant bone and soft tissues.

2. Material and Methods

2.1. Patient Population and Research Design

For the present study, patients aged between 20 to 63 years, that needed replacement of missing teeth in the posterior region of the mandible with adequate condition of remaining bone (height ≥10 mm and width ≥6 mm) and adequate prosthetic space to rehabilitation, were selected in a private clinic (Montevideo, Uruguay). A total of 30 patients, 18 women and 12 men, were consecutively included. All patients signed a written Helsinki informed consent for participation and permission to use the data obtained for research purposes prior to the procedures. The general health condition stability of the participants in the study was considered and their ability to withstand surgery to install the planned implants. Patients with systemic alterations (diabetes, hypertension, or osteoporosis) or local changes (oral pathology in soft or hard tissues, bruxism, and smoking) were excluded from this study.

In addition, patients with uncontrolled and/or untreated periodontal disease, lack of adequate bone tissue for implant insertion, and/or presence of inflammatory events were not included in the study.

Sixty dental implants of conical macro design with cone Morse connection manufactured in grade IV titanium (Implacil De Bortoli, São Paulo, Brazil) were used in the implantations. The implants dimensions used were 3.5 and 4 mm in diameter and 7, 9, and 11 mm in length. Moreover, five implants were used to the surface analysis. The Figure 1 show a representative image of the macro design and the connection characteristics of the implant used in the present study.

Figure 1. Representative image of the macro design and the connection characteristics of the implants used in the study, respectively.

2.2. *Dental Implant Surface Topography*

All implants are treated with sandblasted acid-etched surface technology as previously described by Gehrke et al. [21]. The implants were blasted with 50–100 μm TiO_2 microparticles and, following, the surface was ultrasonically cleaned with an alkaline solution, washed in distilled water, and pickled with maleic acid ($HO_2CCHCHCO_2H$). After these treatments, five implants were used to evaluate the surface characteristics by scanning electron microscopy (SEM, model JSM 5200, JEOL Ltd., Tokyo, Japan) and the roughness parameters, which was measured on the profilometer (Perthometer S2, Mahr GmbH, Göttingen, Germany), where Ra is the absolute value of all profile points, and Rz is the value of the absolute heights of the five highest peaks and the depths of the five deepest valleys.

2.3. *Surgical Procedure of Implant Placement*

All procedures (pre, trans, and postoperative) were performed by two specialists in implantology (MB and JGA). In all patients, mucosal thickness was measured at the local determined for implant installation through a periapical X-ray images. The measurement of the mesio-distal diameter of clinical crown of the tooth adjacent to the place where the implant will be installed was used to calibrate the program. The surgical procedures routinely used to install dental implants were applied. Surgical guides were prepared and used for the installation of all implants. All patients were given antibiotic premedication that was administered orally (2 g of Amoxicillin, 2 h before surgery) and continued in the postoperative for another five days (500 mg every 8 h). After the application of local anesthesia using Articaine 2% (DFL Ltd.a, Rio de Janeiro, Brazil), an incision was performed in the central area of the mucosal crest, and only the buccal flap was displaced, keeping the lingual mucosa in position for the proper measurement of its clinical thickness. The mucosa thickness was measured using a periodontal probe (Hu-Friedy, Chicago, IL, USA), from the crestal bone to the more apical area of the mucosa. Then, the lingual flap was raised, and the implant procedures were performed in accordance to the manufacturer's instructions.

The dimensions of the implants were previously determined during the planning of each case. For osteotomies, a Driller BLM600 motor and a counter angle with a 20:1 reduction (Driller, São Paulo, Brazil) was used under intense external irrigation with 0.9% saline solution. All implants were positioned 2 ± 0.2 mm subcrestally. All sutures were performed using simple point with Nylon 5-0 (Ethicon US, Bridgewater, NJ, USA). For the post-operative pain and inflammation control was administrated Cetoprofeno (200 mg/day) for four days plus paracetamol (750 mg, in case of pain). Ninety days after performing the surgery for the installation of the implants, the rehabilitation procedures were started.

2.4. Clinical Stability and Radiographic Evaluations

The stability of all implants was evaluated by resonance frequency immediately after the installation (t1) and 90 days (t2) in the reentry surgery to install the healing abutment. This evaluation was performed using the Ostell™ Mentor (Integration Diagnostics AB, Goteborg, Sweden) plus the Smartpeg™ (Integration Diagnostics AB) devices. Smartpeg sensors were installed in each implant using a controlled torque of 10 Ncm, as recommended by a recent study [22]. A mean was performed with the values obtained in the measurements in the vestibule-lingual direction (V-L) and mesio-distal direction (M-D).

Three periapical radiographies were made for each patient to measure the mucosa thickness (before implant placement) and the marginal bone loss (immediately after the surgery and 90 days later). Parallel profile radiography using a digital ring holder was used to standardize the analysis and decrease the image distortions. The relation between the implant platform and the crestal bone position was measured. All radiographic measurements were performed using the ImageJ software for Windows (developed at the U.S. National Institutes of Health and available at http://rsb.info.nih.gov/ij). In the first radiography the mucosa width was measured and compared with the clinical measurements. Then, the implants were grouped according to the measured thickness of the mucosa in each implant: Patients with mucosal thickness (MT) between 1.0 and 2.0 mm (MT1); mucosal thickness between 2.1 and 3.0 mm (MT2); and, mucosal thickness more 3.1 mm (MT3). In the radiographs post-implantation, the cortical bone level to the platform was measured and recorded at the mesial-marginal bone loss (m-MBL) and distal-marginal bone loss (d-MBL) side of each implant.

2.5. Statistical Analyses

The methodology and statistical data analyses were reviewed by an independent statistician. The outcomes were longitudinally analyzed among the two initial stability of the implants (ISQ) tests and the bone level between the variables using the one-way analysis of variance (ANOVA) test for repeated measures. The comparison between the clinical and radiographic measurements was performed using the z-test for unpaired samples. The Kolmogorov-Smirnov normality test and the Levene's homogeneity of variance test were used. For bivariate analysis, Mann-Whitney U, and Students-t tests were used. Repeated-measures ANOVA was used to analyze the reduction in marginal bone loss. All comparison analysis was performed using GraphPad Prism 5 software for Windows (GraphPad Software, San Diego, CA, USA). The level of significance was set at $\alpha = 0.05$.

3. Results

The mean and standard deviation of the absolute values of all profile points (Ra) was 0.87 ± 0.14 µm, the root-mean-square of the values of all points (Rq) was 1.12 ± 0.18 µm, and the average value of the absolute heights of the five highest peaks and the depths of the five deepest valleys (Rz) was 5.14 ± 0.69 µm. In the Figure 2 are showed the implant macro design and the surface images of the morphology.

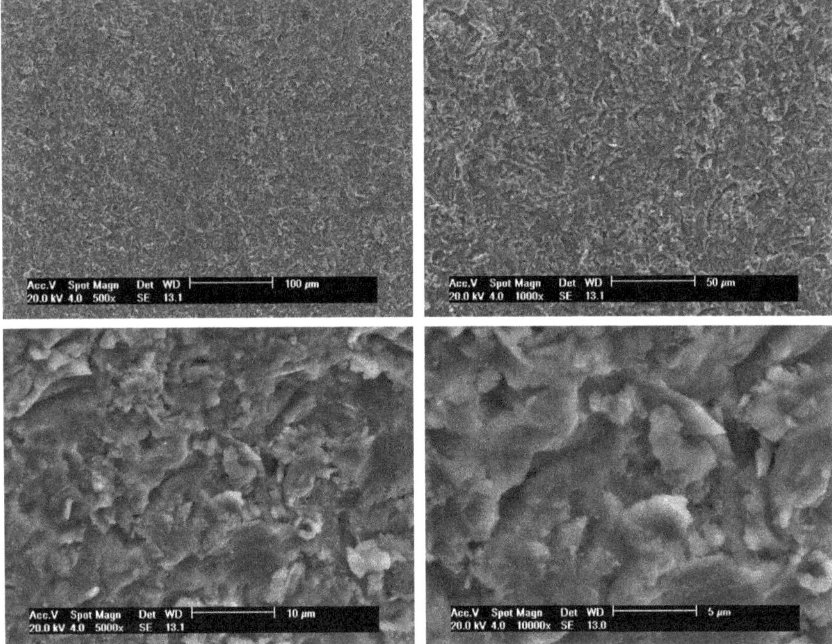

Figure 2. SEM images of the implant surface in different increases.

A total of 60 conical implants of Morse taper connections were installed and the evaluated variables were the diameter of 3.5 mm (n = 17) and 4 mm (n = 43), the different lengths that ranged from 7 mm (n = 21), 9 mm (n = 24), and 11 mm (n = 15) and the mucosa thickness MT1 (n = 19), MT2 (n = 24), and MT3 (n = 17). Thirty patients (18 women and 12 men; ages from 20 to 63 years) received dental implants. After the initial period of 90 days, only one implant was loose throughout the study period and re-implanted with success. Then, the analysis was performed with a total implant quantity (60 implants). No patient dropout was observed during the observation period.

The analyses between the radiographic and clinical measurements showed similar values for the mucosal thickness, no presenting statistical differences ($p = 0.634$), with a mean and standard deviation of 2.26 ± 0.72 and 2.34 ± 1.27, for clinical and radiographic measurements, respectively. The Figure 3 show a box plots graph to compare the measured values and the Figure 4 shows the clinical and radiographic representative image of these measurements.

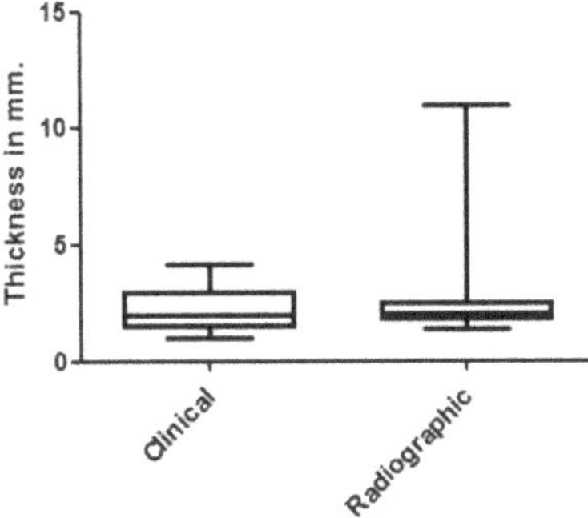

Figure 3. Box-plots graph of the values measured clinical and radiographically.

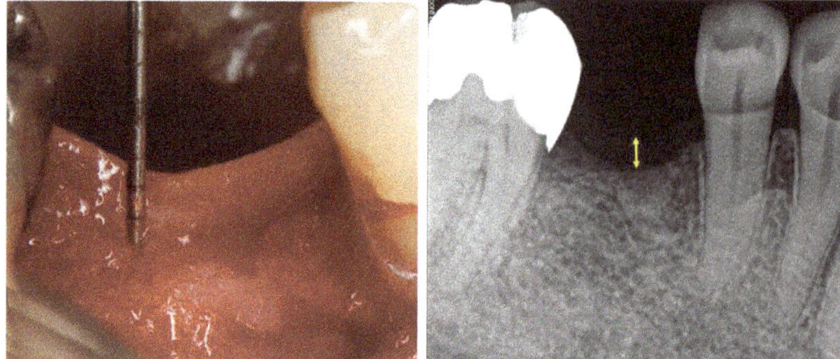

Figure 4. Representative clinical image of the mucosal thickness measurement after the mucosal flap and a periapical x-ray image of the mucosal measurement.

The measured ISQ values showed an overall mean and standard deviation in each proposed time as following: In time t1 was 63.6 ± 2.90 (95% CI: 53 to 70) and in t2 was 69.0 ± 4.14 (95% CI: 49 to 75). The values (mean, SD, and median) are summarized in the Table 1. Figure 5 showed a box-plots graph of the ISQ evolution in each time. No statistical difference of ISQ was observed regarding the implant diameter and length ($p > 0.05$). However, comparing the values of t1 versus t2, an expected statistical difference was found ($p < 0.0001$).

Table 1. Initial stability of the implants (ISQ) analysis and measurements at initial day (baseline) and 90 days after the implant installation. Results as mean and medians.

ISQ Value	Baseline		90 Days	
	Mean ± SD	Median	Mean ± SD	Median
Mesio-distal	63.4 ± 2.94	63.75	68.0 ± 3.85	69.53
Vestibule-lingual	63.7 ± 2.89	64.45	69.9 ± 4.14	70.80

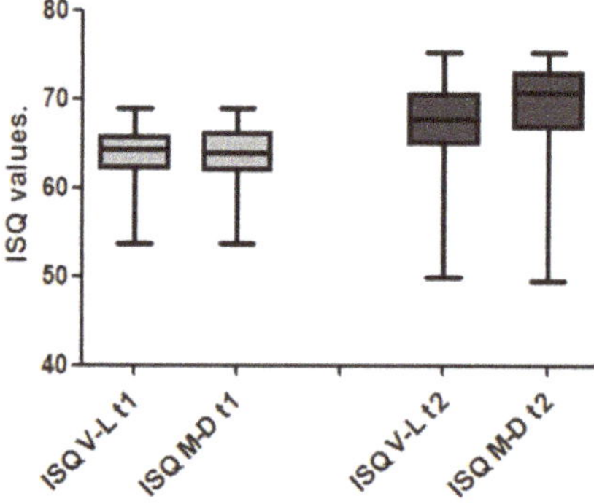

Figure 5. Box-plots graph of the ISQ measured values in the time 1 (t1) and time 2 (t2). V-L = vestibule-lingual direction and M-D = mesio-distal direction.

The comparative data measured between mesial and distal marginal bone loss with the observed variables. Overall mean of mesial and distal MBL were 1.11 ± 1.16 mm and 1.11 ± 1.15 mm, respectively, resulted in non-statistically significant differences ($p > 0.05$). The comparison of the bone loss between the patient's sex showed non-statistically significant differences ($p > 0.05$), where the woman patient's show an MBL mean value of 1.1 ± 1.25 mm and the man patient´s 1.0 ± 0.93 mm. The images of the Figure 6 show a sequence of measurements of the MBL.

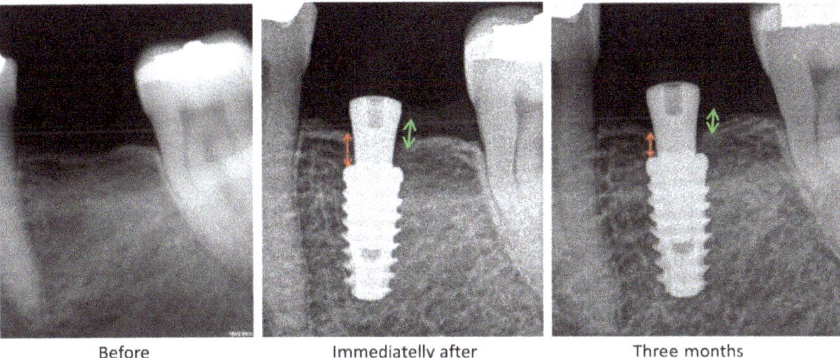

Figure 6. Radiograph sequence used to evaluate and measure the bone level. The measurements were performed from the implant platform to the crestal bone (red arrows = m-MBL and green arrows = d-MBL).

Regarding the implant dimensions, the diameter showed a mean value of MBL in 0.73 ± 0.8 mm for the implants of 3.5 mm and 1.05 ± 1.1 mm for the implant of 4.0 mm, with significant statistical difference ($p < 0.001$). The bar graph of the Figure 7 shows the values of mesial and distal MBL measurements.

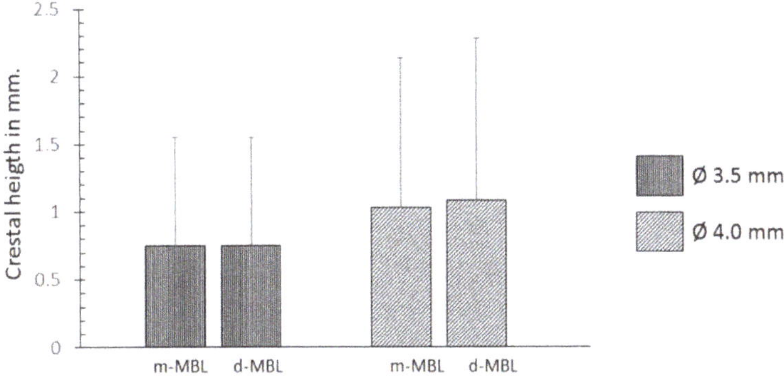

Figure 7. Bar graph showing the mean and standard deviation of the MBL values measured in mesial (m-MBL) and distal (d-MBL) position of each implant in the two implant diameters.

However, the MBL values measured at different implant lengths showed very similar values (Figure 8), without statistically significant differences ($p > 0.05$).

The mucosal thickness (MT1, MT2, and MT3) resulted in statistically significant differences ($p < 0.05$). The mean value of MBL in the MT1 was 1.5 ± 0.8 mm, in the MT2 was 0.75 ± 0.5 and in the MT3 was 0.9 ± 0.8 mm. The bar graph of the Figure 9 shows the values for mesial and distal measurements. In general, the better behavior was observed in the MT2 (mucosal thickness between 2.1 and 3.0 mm), with 0.7 ± 0.6 mm for m-MBL and 0.8 ± 0.5 mm for d-MBL.

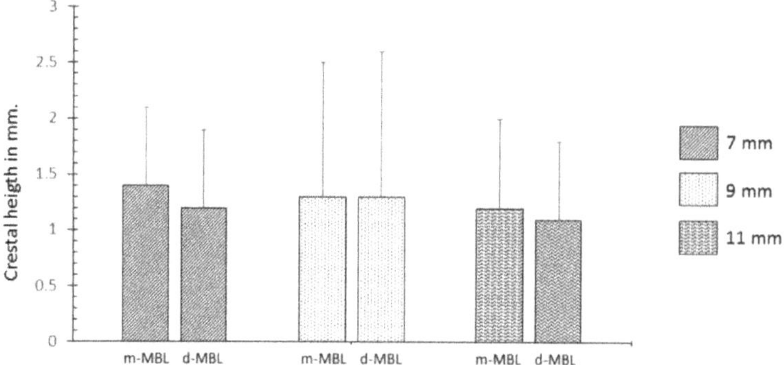

Figure 8. Bar graph showing the mean and standard deviation of the MBL values measured in mesial (m-MBL) and distal (d-MBL) position of each implant in the three implant lengths used.

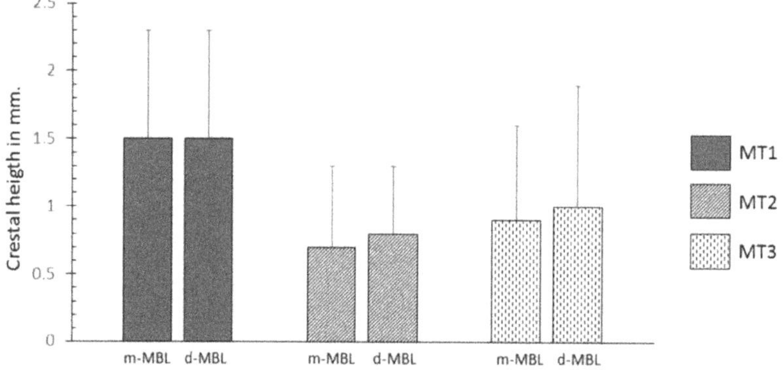

Figure 9. Bar graph showing the mean and standard deviation of the MBL values measured in mesial (m-MBL) and distal (d-MBL) position of each implant in the three mucosal thickness.

4. Discussion

This clinical study describes an analysis the marginal bone behavior considering of different variables after the implants installed in the posterior area of mandible: Patient sex, implant dimensions (diameter and length), and mucosa thickness. One implant was loose throughout the study period and re-implanted, and the survival rate of dental implants considered was of 98.3%. This study was made without patient selection, the only criterion was the posterior inferior region selection, this makes a random distribution of the ridges with the difficulty of achieving uniformity in the height and width of the ridge and the trouble of achieving the 2 mm of supracrestal bone in all cases, as well as having bone in around the 100% of the perimeter, simply because of the anatomy of the area. These could be factors that may affect the mucosal position and the behavior around the implant.

All implants measured the mucosa thickness on the radiographic images and compared with the clinical measurements, and the results confirmed no statistical differences among these collected data. Then, the measurement of the mucosa in radiographic images can be used for planification of the implant position (depth position). Therefore, the ISQ and the bone height in relation of the implant platform were measured immediately after the implant placement (baseline) and after 90 days. The relation of the marginal bone behavior and the variables with significant statistical differences are discussed separately follow.

Surface topography refers to the degree of surface roughness and the orientation of surface irregularities, which can directly stimulate osseointegration, increasing and/or accelerating the events involved in this process [8,21]. In this sense, we use implants with a surface roughness considered moderate, similar to that used by many other brands of implants. This data is important so that in future studies researchers can compare their results or reproduce in new investigations. In addition, since stability measurements were taken 90 days after implantation, the results are directly affected by the type of surface used in the implants.

4.1. Initial Implant Stability

The initial stability of the implants, a measure that can be represented by ISQ, is of fundamental importance for osseointegration. Several studies describe a direct relationship between bone density and measured ISQ values [23–27]. Both the thickness of the cortical bone and the pattern presented by the medullary portion (trabecular), which are in contact with the installed implant, are determinant factors for stability (bone and implant contact) [28]. The aim of this study was to observe the consequences of the placing Morse taper implants subcrestally and the relation between bone remodeling and soft tissue thickness. The initial results in these three months of studies show that the sectors in which the mucosal thickness was 1 to 2 mm suffered greater bone remodelation compared to the sectors where the width of the soft tissue was 2 mm or more.

The clinical methods that are commonly used to verify implant stability and osseointegration include percussion, mobility, and radiographic studies. However, these methods have an important limitation in their standardization, since they have a great dependence on the sensitivity and susceptibility with respect to the professional executor [29,30]. In this sense, more precise and non-invasive techniques were developed. These analyzes are called according to the method by which they are performed, i.e., resonance frequency analysis (RFA), and are used to verify and measure the stability of implants installed in bone tissue at different clinical periods [30,31]. The use of this technique is mainly based on being easy to perform, fast, and direct and, moreover, can be applied routinely in the clinic because it does not present discomfort to the patient.

The measured values of ISQ varied during the phases of osseointegration evaluated. At the trans-operative time, the mean and standard deviations of the ISQ values measured was 63.6 ± 2.90 varying of 53 to 70, indicating adequate primary stability, similar to the results reported in other studies that presented averages from 60.3 to 62.6 [31–34]. While, in the second time measured, 90 days after the implantations, the means and standard deviations of the ISQ values was 69.7 ± 7.09 varying of 49 to 75. Such overall result for the ISQ values for the time of 90 days are within the mean values shown in several similar studies, where the values varied from 67.0 to 72.1 [31,32,35,36].

4.2. Mucosal Thickness

Linkevicius and colleagues studied the main factors related with the bone remodeling [2,3], understanding and trying to make a relation between mechanic factors and mucosal thickness. Isolating the connection factor even when this was 2 mm supracrestal and the mucosa was thin, there was bone remodeling consequence of the biological width formation [2]. Using implants with platform switching concept did not prevent the bone remodeling also when the mucosal thickness shows little thickness [3,4]. The use of Morse-cone implants for this study is based on the minimal number of microorganisms penetrating the implant/abutment microgap as well as the absence of movement [37]. This rigid type of connection eliminates a potential remodeling bone factor and with the subcrestal position opens a new way in which the biologic width can be conformed [37–40]. Studies showed the possibility of having no bone remodeling reaction at the abutment-implant interface and making a new configuration of the biological space and the mucosa characteristics surrounding the implant [41–44].

The results are in agreement with the studies mentioned previously, although in our study the evaluation time was less than one year, the first case (MT1), where the mucosa was between 1 and 2 mm, suffered much more bone loss than de MT2 and MT3 where the mucosal width was two or more millimeters. The group MT2 and MT3 had very similar measures in respect of bone remodeling. However, the MT3 not show superior behavior in comparison with the MT1 and MT2, possibly because in this group the biological space exceeded the value considered ideal for the position of implant-abutment junction (IAJ). Conversely, controversial information is available regarding implants placed subcrestally. Some authors recommended placement of the implant platform 1 or 2 mm below the alveolar crest to better maintain marginal bone levels [45,46]. However, other studies reported an increased extension of inflammatory infiltrate due to deep positioning of the IAJ, resulting in greater MBL compared to implants placed equicrestally [47,48]. In the case of implants with mucosal thickness 3 mm (MT3), added to 2 mm positioning subcrestal implants, the final positioning distance of the AIJ was greater than 5 mm, which probably explains the behavior of the implants in this condition.

Clinically the time bone loss around implants can influence the planned aesthetic results, mainly because it alters the final positioning of the tissue because its final volume decreased.

4.3. Implant Dimensions (Diameter and Length)

Several studies showed that the implant dimensions have direct influence on the stress distribution to the bone [41–45] as the implant directly affects the area of possible bone retention [49]. Moreover, other authors have advocated the use of implants as long and wide as possible [50]. However, when bone loss around the implants was evaluated, there is a controversy regarding the influence of length in these alterations, and some studies present results of larger losses in the short implants, other authors report that the length of the implant has little influence on the quantity of vertical load stress, and may have a lower effect on the distribution of stresses to bone tissue when compared to the variation in implant diameter [49,51]. In this way, Koutouzis and collaborates not found statistical difference in the values of bone resorption in larger diameter implants, comparing small diameter (3.5 mm) with larger diameter (4.5 mm) [52]. Unlike most of the cited reports, an important fact that the research revealed was the better behavior of the 3.5 mm implants when compared to the 4.0 mm diameter implants in terms of bone remodeling, that is, the smaller implants diameters showed lower bone resorption. On the other hand, when the implants were compared in terms of length (7, 9, and 11 mm), the results obtained in the present study did not present statistical differences, with very similar values among the sizes used. However, in the present study the implants evaluated were not placed under masticatory loads, which may be the reason for the difference in results between the studies. Previous FEA studies have shown that a decrease in diameter increases the stress transferred to crestal bone [53].

5. Conclusions

Within the limitations of this prospective study, although in our study the evaluation time was short after the implantation, we concluded that cone Morse implants placed 2 mm subcrestal level showed different values of bone loss depending of the mucosal thickness and implant diameter. However, the initial stability and implant length not showed influence on the marginal bone loss.

Author Contributions: Conceptualization, J.C.P.-F.; Data curation, S.A.G., M.B. and J.A.; Formal analysis, S.A.G. and J.C.P.-F.; Investigation, M.B. and J.A.; Methodology, J.L.C.-G., S.A.G. and J.C.P.-F.; Project administration, M.B., J.L.C.-G.; Resources, J.A.; Software, M.B. and J.A.; Supervision, S.A.G. and J.C.P.-F.; Visualization, J.C.P.-F.; Writing—original draft, M.B. and J.A.; Writing—review and editing, J.L.C.-G. and S.A.G.

Funding: This research received no external funding.

Acknowledgments: The authors are grateful to the Paulo Rossetti from the Dental School of the University of São Paulo, Bauru, Brazil, for their kind support as independent statistician. The authors are grateful to Implacil De Bortoli (São Paulo, Brazil) for the products support.

Conflicts of Interest: The authors declare no conflict of interest.

References

1. Berglundh, T.; Lindhe, J. Dimension of the periimplant mucosa. Biological width revisited. *J. Clin. Periodontol.* **1996**, *23*, 971–973. [CrossRef] [PubMed]
2. Linkevicius, T.; Apse, P.; Grybauskas, S.; Puisys, A. The influence of soft tissue thickness on crestal bone changes around implants: A 1-year prospective controlled clinical trial. *Int. J. Oral Maxillofac. Implant.* **2009**, *24*, 712–719.
3. Linkevicius, T.; Apse, P.; Grybauskas, S.; Puisys, A. Influence of thin mucosal tissues on crestal bone stability around implants with platform switching: A 1-year pilot study. *J. Oral Maxillofac. Surg.* **2010**, *68*, 2272–2277. [CrossRef] [PubMed]
4. van Eekeren, P.; Tahmaseb, A.; Wismeijer, D. Crestal bone changes in macrogeometrically similar implants with the implant-abutment connection at the crestal bone level or 2.5 mm above: A prospective randomized clinical trial. *Clin. Oral Implant. Res.* **2016**, *27*, 1479–1484. [CrossRef] [PubMed]
5. Linkevicius, T.; Puisys, A.; Svediene, O.; Linkevicius, R.; Linkeviciene, L. Radiological comparison of laser-microtextured and platform-switched implants in thin mucosal biotype. *Clin. Oral Implant. Res.* **2015**, *26*, 599–605. [CrossRef]
6. Puisys, A.; Linkevicius, T. The influence of mucosal tissue thickening on crestal bone stability around bone-level implants. A prospective controlled clinical trial. *Clin. Oral Implant. Res.* **2015**, *26*, 123–129. [CrossRef]
7. Linkevicius, T.; Puisys, A.; Steigmann, M.; Vindasiute, E.; Linkeviciene, L. Influence of Vertical Soft Tissue Thickness on Crestal Bone Changes Around Implants with Platform Switching: A Comparative Clinical Study. *Clin. Implant Dent. Relat. Res.* **2015**, *17*, 1228–1236. [CrossRef]
8. Gehrke, S.A.; da Silva Neto, U.T. Evaluation of the Surface Treatment on Bone Healing in a Transmucosal 1-mm Area of Implant Abutment: An Experimental Study in the Rabbit Tibia. *Clin. Implant Dent. Relat. Res.* **2016**, *18*, 489–497. [CrossRef]
9. van Eekeren, P.J.; Tahmaseb, A.; Wismeijer, D. Crestal Bone Changes Around Implants with Implant-Abutment Connections at Epicrestal Level or Above: Systematic Review and Meta-Analysis. *Int. J. Oral Maxillofac. Implant.* **2016**, *31*, 119–124. [CrossRef]
10. Schwarz, F.; Hegewald, A.; Becker, J. Impact of implant-abutment connection and positioning of the machined collar/microgap on crestal bone level changes: A systematic review. *Clin. Oral Implant. Res.* **2014**, *25*, 417–425. [CrossRef]
11. Hermann, J.S.; Schoolfield, J.D.; Schenk, R.K.; Buser, D.; Cochran, D.L. Influence of the size of the microgap on crestal bone changes around titanium implants. A histometric evaluation of unloaded non-submerged implants in the canine mandible. *J. Periodontol.* **2001**, *72*, 1372–1383. [CrossRef] [PubMed]
12. Zanatta, L.C.; Dib, L.L.; Gehrke, S.A. Photoelastic stress analysis surrounding different implant designs under simulated static loading. *J. Craniofacial Surg.* **2014**, *25*, 1068–1071. [CrossRef] [PubMed]
13. Stein, A.E.; McGlmphy, E.A.; Johnston, W.M.; Larsen, P.E. Effects of implant design and surface roughness on crestal bone and soft tissue levels in the esthetic zone. *Int. J. Oral Maxillofac. Implant.* **2009**, *24*, 910–919.
14. Schmitt, C.M.; Nogueira-Filho, G.; Tenenbaum, H.C.; Lai, J.Y.; Brito, C.; Döring, H.; Nonhoff, J. Performance of conical abutment (Morse Taper) connection implants: A systematic review. *J. Biomed. Mater. Res. Part A* **2014**, *102*, 552–574. [CrossRef] [PubMed]
15. Baj, A.; Bolzoni, A.; Russillo, A.; Lauritano, D.; Palmieri, A.; Cura, F.; Silvestre, F.J.; Giannì, A.B. Cone-morse implant connection system significantly reduces bacterial leakage between implant and abutment: An in vitro study. *J. Biol. Regul. Homeost. Agents* **2017**, *31*, 203–208. [PubMed]
16. Gehrke, S.A.; Delgado-Ruiz, R.A.; Prados Frutos, J.C.; Prados-Privado, M.; Dedavid, B.A.; Granero Marín, J.M.; Calvo Guirado, J.L. Misfit of Three Different Implant-Abutment Connections Before and After Cyclic Load Application: An in Vitro Study. *Int. J. Oral Maxillofac. Implant.* **2017**, *32*, 822–829. [CrossRef] [PubMed]
17. Hanaoka, M.; Gehrke, S.A.; Mardegan, F.; Gennari, C.R.; Taschieri, S.; Del Fabbro, M.; Corbella, S. Influence of implant/abutment connection on stress distribution to implant-surrounding bone: A finite element analysis. *J. Prosthodont.* **2014**, *23*, 565–571. [CrossRef] [PubMed]
18. Albrektsson, T.; Zarb, G.; Worthington, P.; Eriksson, A.R. The long-term efficacy of currently used dental implants: A review and proposed criteria of success. *Int. J. Oral Maxillofac. Implant.* **1986**, *1*, 11–25.

19. Albrektsson, T.; Chrcanovic, B.; Östman, P.O.; Sennerby, L. Initial and long-term crestal bone responses to modern dental implants. *Periodontology 2000* **2017**, *73*, 41–50. [CrossRef]
20. Sasada, Y.; Cochran, D.L. Implant-Abutment Connections: A Review of Biologic Consequences and Peri-implantitis Implications. *Int. J. Oral Maxillofac. Implant.* **2017**, *32*, 1296–1307. [CrossRef]
21. Gehrke, S.A.; Ramírez-Fernandez, M.P.; Granero Marín, J.M.; Barbosa Salles, M.; Del Fabbro, M.; Calvo Guirado, J.L. A comparative evaluation between aluminium and titanium dioxide microparticles for blasting the surface titanium dental implants: An experimental study in rabbits. *Clin. Oral Implant. Res.* **2018**, *29*, 802–807. [CrossRef] [PubMed]
22. Salatti, D.B.; Pelegrine, A.A.; Gehrke, S.; Teixeira, M.L.; Moshaverinia, A.; Moy, P.K. Is there a need for standardization of tightening force used to connect the transducer for resonance frequency analysis in determining implant stability? *Int. J. Oral Maxillofac. Implant.* **2019**, *34*, 886–890. [CrossRef] [PubMed]
23. Balleri, P.; Cozzolino, A.; Ghelli, L.; Momichioli, G.; Varriale, A. Stability measurements of osseointegrated implants using Osstell in partially edentulous jaws after one year of loading: A pilot study. *Clin. Implant. Dent. Relat. Res.* **2002**, *4*, 128–132. [CrossRef] [PubMed]
24. Barewal, R.M.; Oates, T.W.; Meredith, N.; Cochran, D.L. Resonance frequency measurement of implant stability in vivo on implants with a sandblasted and acid-etched surface. *Int. J. Oral Maxillofac. Implant.* **2003**, *18*, 641–651.
25. Bischof, M.; Nedir, R.; Szmukler-Moncler, S.; Bernard, J.P.; Samson, J. Implant stability measurement of delayed and immediately loaded implants during healing. *Clin. Oral Implant. Res.* **2004**, *15*, 529–539. [CrossRef] [PubMed]
26. Nedir, R.; Bischof, M.; Szmulzkler-Moncler, S.; Bernard, J.P.; Samson, J. Predicting osseointegration by means of implant primary stability: A resonance frequency analysis with delayed and immediately loaded ITI SLA implants. *Clin. Oral Implant. Res.* **2004**, *15*, 520–528. [CrossRef] [PubMed]
27. Oates, T.W.; Valderrama, P.; Bischof, M.; Nedir, R.; Jones, A.; Simpson, J.; Toutenburg, H.; Cochran, D.L. Enhanced implant stability with a chemically modified SLA surface: A randomized pilot study. *Int. Oral Maxillofac. Implant.* **2007**, *22*, 755–760.
28. Meredith, N. Assessment of implant stability as a prognostic determinant. *Int. J. Prosthodont.* **1998**, *11*, 491–501.
29. Fischer, K.; Bäckström, M.; Sennerby, L. Immediate and early loading of oxidized tapered implants in the partially edentulous maxilla: A 1-year prospective clinical, radiographic, and resonance frequency analysis study. *Clin. Implant. Dent. Relat. Res.* **2009**, *11*, 69–80. [CrossRef]
30. Meredith, N.; Book, K.; Friberg, B.; Jemt, T.; Sennerby, L. Resonance frequency measurements of implant stability in vivo. A crosssectional and longitudinal study of resonance frequency measurements on implants in the edentulous and partially dentate maxilla. *Clin. Oral Implant. Res.* **1997**, *8*, 226–233. [CrossRef]
31. Gehrke, S.A.; da Silva Neto, U.T.; Rossetti, P.H.; Watinaga, S.E.; Giro, G.; Shibli, J.A. Stability of implants placed in fresh sockets versus healed alveolar sites: Early findings. *Clin. Oral Implant. Res.* **2016**, *27*, 577–582. [CrossRef] [PubMed]
32. Gehrke, S.A.; Tavares da Silva Neto, U. Does the time of osseointegration in the maxilla and mandible differ? *J. Craniofacial Surg.* **2014**, *25*, 2117–2120. [CrossRef] [PubMed]
33. Friberg, B.; Sennerby, L.; Linden, B.; Gröndahl, K.; Lekholm, U. Stability measurements of one-stage Branemark implants during healing in mandibles. A clinical resonance frequency analysis study. *Int. J. Oral Maxillofac. Surg.* **1999**, *28*, 266–272. [CrossRef]
34. Zix, J.; Hug, S.; Kessler-Liechti, G.; Mericske-Stern, R. Measurement of dental implant stability by resonance frequency analysis and damping capacity assessment: Comparison of both techniques in a clinical trial. *Int. J. Oral Maxillofac. Implant.* **2008**, *23*, 525–530.
35. da Silva Neto, U.T.; Joly, J.C.; Gehrke, S.A. Clinical analysis of the stability of dental implants after preparation of the site by conventional drilling or piezosurgery. *Br. J. Oral Maxillofac. Surg.* **2014**, *52*, 149–153. [CrossRef] [PubMed]
36. Gehrke, S.A.; da Silva, U.T.; Del Fabbro, M. Does Implant Design Affect Implant Primary Stability? A Resonance Frequency Analysis-Based Randomized Split-Mouth Clinical Trial. *J. Oral Implantol.* **2015**, *41*, e281–e286. [CrossRef]

37. Koutouzis, T.; Neiva, R.; Nair, M.; Nonhoff, J.; Lundgren, T. Cone beam computed tomographic evaluation of implants with platform-switched Morse taper connection with the implant-abutment interface at different levels in relation to the alveolar crest. *Int. J. Oral Maxillofac. Implant.* **2014**, *29*, 1157–1163. [CrossRef] [PubMed]
38. Koutouzis, T.; Neiva, R.; Nonhoff, J.; Lundgren, T. Placement of implants with platform-switched Morse taper connections with the implant-abutment interface at different levels in relation to the alveolar crest: A short-term (1-year) randomized prospective controlled clinical trial. *Int. J. Oral Maxillofac. Implant.* **2013**, *28*, 1553–1563. [CrossRef]
39. Fetner, M.; Fetner, A.; Koutouzis, T.; Clozza, E.; Tovar, N.; Sarendranath, A.; Coelho, P.G.; Neiva, K.; Janal, M.N.; Neiva, R. The Effects of Subcrestal Implant Placement on Crestal Bone Levels and Bone-to-Abutment Contact: A Microcomputed Tomographic and Histologic Study in Dogs. *Int. J. Oral Maxillofac. Implant.* **2015**, *30*, 1068–1075. [CrossRef]
40. Tesmer, M.; Wallet, S.; Koutouzis, T.; Lundgren, T. Bacterial colonization of the dental implant fixture-abutment interface: An in vitro study. *J. Periodontol.* **2009**, *80*, 1991–1997. [CrossRef]
41. Welander, M.; Abrahamsson, I.; Berglundh, T. Subcrestal placement of two-part implants. *Clin. Oral Implant. Res.* **2009**, *20*, 226–231. [CrossRef] [PubMed]
42. Weng, D.; Nagata, M.J.H.; Bell, M.; Bosco, A.F.; de Melo, L.G.N.; Richter, E.-J. Influence of microgap location and configuration on the periimplant bone morphology in submerged implants. An experimental study in dogs. *Clin. Oral Implant. Res.* **2008**, *19*, 1141–1147. [CrossRef] [PubMed]
43. Weng, D.; Nagata, M.J.H.; Leite, C.M.; de Melo, L.G.N.; Bosco, A.F. Influence of microgap location and configuration on radiographic bone loss in nonsubmerged implants: An experimental study in dogs. *Int. J. Prosthodont.* **2010**, *24*, 445–452.
44. Pilliar, R.M.; Deporter, D.A.; Watson, P.A.; Valiquette, N. Dental implant design: Effect on bone remodeling. *J. Biomed. Mater. Res.* **1991**, *25*, 467–483. [CrossRef] [PubMed]
45. Brunski, J.B. Biomechanical considerations in dental implant design. *Int. J. Oral Maxillofac. Implant.* **1988**, *5*, 31–34.
46. Holmgren, E.P.; Seckinger, R.J.; Kilgren, L.M.; Mante, F. Evaluating parameters of osseointegrated dental implants using finite element analysis—A two-dimensional comparative study examining the effects of implant diameter, implant shape, and load direction. *J. Oral Maxillofac. Implant.* **1998**, *24*, 80–88. [CrossRef]
47. Meijer, H.J.; Kuiper, J.H.; Starmans, F.J.; Bosman, F. Stress distribution around dental implants: Influence of superstructure, length of implants, and height of mandible. *J. Prosthet. Dent.* **1992**, *68*, 96–102. [CrossRef]
48. Iplikcioglu, H.; Akca, K. Comparative evaluation of the effect of diameter, length and number of implants supporting three-unit fixed partial prostheses on stress distribution in the bone. *J. Dent.* **2002**, *30*, 41–46. [CrossRef]
49. Himmlova, L.; Dostalova, T.; Kacovsky, A.; Konvickova, S. Influence of implant length and diameter on stress distribution: A finite element analysis. *J. Prosthet. Dent.* **2004**, *91*, 20–25. [CrossRef]
50. Winkler, S.; Morris, H.F.; Ochi, S. Implant survival to 36 months as related to length and diameter. *Ann. Periodontol.* **2000**, *5*, 22–31. [CrossRef]
51. Misch, C.E. Divisions of available bone. In *Contemporary Implant Dentistry*, 2nd ed.; Mosby: St Louis, MO, USA, 1999; pp. 91–94.
52. Koutouzis, T.; Fetner, M.; Fetner, A.; Lundgren, T. Retrospective evaluation of crestal bone changes around implants with reduced abutment diameter placed non-submerged and at subcrestal positions: The effect of bone grafting at implant placement. *J. Periodontol.* **2011**, *82*, 234–242. [CrossRef] [PubMed]
53. Qian, L.; Todo, M.; Matsushita, Y.; Koyano, K. Effects of implant diameter, insertion depth, and loading angle on stress/strain fields in implant/ jawbone systems: Finite element analysis. *Int. J. Oral Maxillofac. Implant.* **2009**, *24*, 877–886.

Article

Intraosteal Behavior of Porous Scaffolds: The mCT Raw-Data Analysis as a Tool for Better Understanding

Andrés Parrilla-Almansa [1], Carlos Alberto González-Bermúdez [2], Silvia Sánchez-Sánchez [1], Luis Meseguer-Olmo [3], Carlos Manuel Martínez-Cáceres [4], Francisco Martínez-Martínez [5], José Luis Calvo-Guirado [6], Juan José Piñero de Armas [7], Juan Manuel Aragoneses [8], Nuria García-Carrillo [9] and Piedad N. De Aza [10,*]

1 Image Diagnostic Service, Virgen de la Arrixaca University Hospital, El Palmar, 30120 Murcia, Spain; aparrilla10@gmail.com (A.P.-A.); silviasanchez@gmail.com (S.S.-S.)
2 Faculty of Medicine, Universidad de Murcia, Instituto Murciano de Investigación Biosanitaria Virgen de la Arrixaca (IMIB-Arrixaca), 30.100 Murcia, Spain; cagb1@um.es
3 Department of Orthopaedic Surgery and Trauma, School of Medicine, Lab of Regeneration and Tissue Repair, UCAM-Universidad Catolica San Antonio de Murcia, Guadalupe, 30107 Murcia, Spain; lmeseguer.doc@gmail.com
4 Pathology Unit, Biomedical Research Institute of Murcia (IMIB-Arrixaca-UMU), El Palmar, 30120 Murcia, Spain; cmmarti@um.es
5 Orthopaedic and Trauma Service, Virgen de la Arrixaca University Hospital, El Palmar, 30120 Murcia, Spain; fmtnez@gmail.com
6 Department of Oral Surgery and Implant Dentistry, Faculty of Health Sciences, UCAM- Universidad Católica San Antonio de Murcia, Guadalupe, 30107 Murcia, Spain; jlcalvo@ucam.edu
7 Cátedra Internacional de Análisis Estadístico y Big Data, Universidad Católica de Murcia, 30107 Murcia, Spain; jjpinero@ucam.edu
8 Department of Dental Research in Universidad Federico Henriquez y Carvajal (UFHEC), Santo Domingo 10107, Dominican Republic; jaragoneses@ufhec.edu.do
9 Department of Medicina Oral, Facultad de Medicina, Universidad de Murcia, Instituto Murciano de Investigación Biosanitaria Virgen de la Arrixaca (IMIB-Arrixaca), 30.100 Murcia, Spain; ngc2@um.es
10 Instituto de Bioingenieria, Universidad Miguel Hernandez, 03202 Elche, Spain
* Correspondence: piedad@umh.es

Received: 10 March 2019; Accepted: 10 April 2019; Published: 12 April 2019

Abstract: The aim of the study is to determine the existing correlation between high-resolution 3D imaging technique obtained through Micro Computed Tomography (mCT) and histological-histomorphometric images to determine in vivo bone osteogenic behavior of bioceramic scaffolds. A Ca-Si-P scaffold ceramic doped and non-doped (control) with a natural demineralized bone matrix (DBM) were implanted in rabbit tibias for 1, 3, and 5 months. A progressive disorganization and disintegration of scaffolds and bone neoformation occurs, from the periphery to the center of the implants, without any differences between histomorphometric and radiological analysis. However, significant differences ($p < 0.05$) between DMB-doped and non-doped materials where only detected through mathematical analysis of mCT. In this way, average attenuation coefficient for DMB-doped decreased from 0.99 ± 0.23 Hounsfield Unit (HU) (3 months) to 0.86 ± 0.32 HU (5 months). Average values for non-doped decreased from 0.86 ± 0.25 HU (3 months) to 0.66 ± 0.33 HU. Combination of radiological analysis and mathematical mCT seems to provide an adequate in vivo analysis of bone-implanted biomaterials after surgery, obtaining similar results to the one provided by histomorphometric analysis. Mathematical analysis of Computed Tomography (CT) would allow the conducting of long-term duration in vivo studies, without the need for animal sacrifice, and the subsequent reduction in variability.

Keywords: ceramic scaffolds; demineralized bone matrix; bone regeneration; micro-CT; histomorphometry

1. Introduction

Imaging techniques have contributed to some of the most significant advances in biomedicine, and this trend is accelerating [1]. Development of image-capture techniques has required parallel advances in image processing and characterization, being consolidated as a specific research discipline in biomedicine. Digital image processing and analysis are aimed to develop methods of information extraction from images generated by different capture techniques, as well as to integrate the information obtained [2].

Compare to others, Computed Tomography (CT) offers the best radiographic method for morphological and qualitative analysis of bone and solid structures [3]. The images obtained through CT are the result of a volumetric decomposition of the body in units or voxels. From each voxel, a data set of attenuation coefficients is extracted, being known as "raw data". This volumetric set of raw data is mathematically reconstructed (iterative reconstruction, filtered retro projection algorithms) to obtain a pixelated image [4,5].

Strongly related to the size/number of voxels, and to the resolution of the acquisition system, the reconstruction from a volumetric object in a pixelated and multiplanar image is associated with the loss of data, as well as to the introduction of errors. In this way, despite presenting different attenuation coefficient, data contained in a voxel would contribute to the final pixel conformation [5]. Apart from the previously explained sources of miss information, the interpretation of radiological images is associated with other problems. At this respect, Garland (1949) [6] warned the scientific community about the error inherent in the radiological interpretation. Among the frequent mistakes identified during imaging analysis, perceptual errors or miss evaluation of images with a limited interpretive value are included [7]. Alternatively, cognitive errors of interpretation have been also described as frequent [8]. Work overload and environment factors are also decisive in radiological interpretation as, among others, they determine staff fatigue or eyestrain [9]. To reduce errors and false negative rates of image interpretation, a computer-assisted detection (CAD) program has been developed. CAD is software to analyze digital data set of the images, and estimate the probability of a specific disease based on algorithms developed to identify structural distortions. So far it has been used on mammographic studies but with limited results [10,11].

New strategies in the management of bone lesions include the implantation of porous biomaterial scaffolds. These biomaterials can enhance bone healing capacity and favor new bone formation when it is comprised (for instance: lack of vascularization, infections, lack of mechanical stability, or tissue loss) [12]. Histomorphometry and Micro Computed Tomography (mCT) proved to be dominant imaging techniques in the preclinical evaluation of cortical and medullary bone structure. These techniques enable the measurement and the assessment of in vivo bone formation and resorption. By using the mCT, either the anatomic correlation in 2D or the spatial bone microstructure in 3D can be analyzed [13,14]. Algorithmic studies have been conducted to ensure the greatest correspondence in the correlation between the image obtained by mCT and the one obtained by histology [12,15]. The development of a standardized method based on raw-data analysis would contribute to non-destructively quantify the formation of new bone tissue at the periphery and within the implanted biomaterial, reducing the variability associated with CT image processing and its interpretation.

The aim of the study is to determine the possible role of a global imaging analysis, which includes imaging descriptive analysis, mCT raw-data mathematical analysis, and histomorphometric analysis in order to objectively elucidate the osteogenic behavior observed after implant porous biomaterial scaffolds in tibias from New Zealand rabbits as preclinical model.

2. Materials and Methods

2.1. Porous Scaffold Implants Composition and Fabrication

Porous scaffolds were produced by partial sintering method [16–19]. Laboratory previously synthesized tricalcium phosphate ($Ca_3(PO_4)_2$) [TCP] and dicalcium silicate (Ca_2SiO_4) [C2S] were used as raw materials [20,21]. C2S (45% wt.) and TCP (55% wt.) were attrition-milled in isopropyl media, dried (60°C overnight), isostatically pressed (200 MPa) and heated inside platinum crucibles for 2 h to 1000 °C. Homogeneous bars were obtained after being grounded, pressed and reheated (1300 °C, 24 h). One part of the material was grounded to obtain a coarse fraction powder of ~1–2 mm particle size, whereas the other part was milled to ~2 μm. Scaffold implants were obtained after mixing 90% of 1–2 mm coarse fraction powder with 10% of 2 μm particles, using polyvinyl acetate (10%) as binder. Mixture was then heated to 1170° for 2 h, allowing cooling to room temperature inside furnace for 24 h. Finally, cylinders (height = 6 mm; diameter = 4.5 mm) were cut and sterilized by means of hydrogen peroxide gas-plasma (Sterrad® 100S, Germany) at low temperature (Figure 1A).

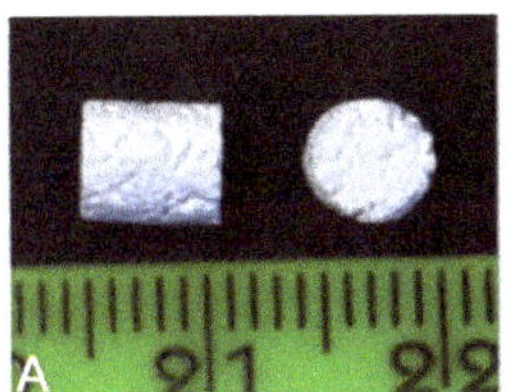

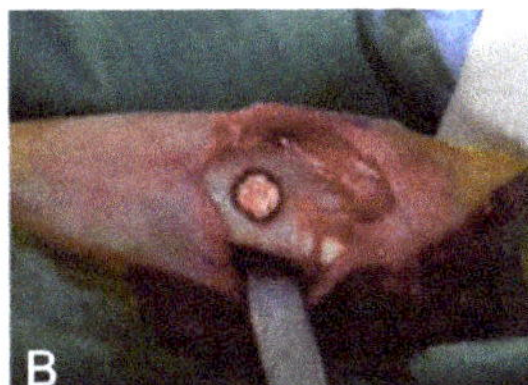

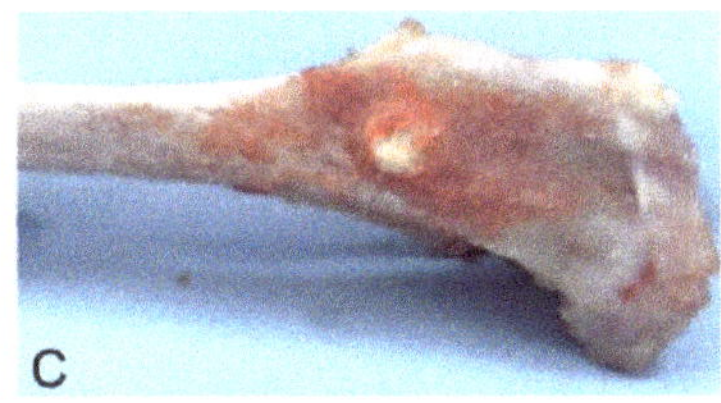

Figure 1. (**A**) Non-doped porous scaffold implants (before being implanted); (**B**) Intrasurgical image (proximal anteromedial metaphysis of right tibia) showing the implanted material; (**C**) Sample collection after tibia removal.

After sterilization, 60% of scaffolds were immersed in 231.3 ± 1.35 mg of Demineralized Bone Matrix Gel (DBM-gel, labeled as DMB-doped)) Activagen® (Bioteck, Arcugnano, Italy) for 3 minutes. Each scaffold was weighed to ensure an impregnated amount of DBM of 79.06 ± 0.47 mg. DBM is being defined as osteogenic inductor, promoting a rapid vascularization and osteoblast differentiation [22]. The remaining 40% of pellets were untreated (labeled as non-doped).

2.2. *Animal Selection and Conditioning*

Fifteen New Zealand white rabbits were included in this study. Inclusion criteria were: pathogen-free male, with an initial weight of 3.500–4.000 g and aged 26–28 weeks to ensure skeletal maturity and physical closure [23,24]. Ethical approval for the experiment was obtained from the Ethics Committee in Animal Research of the University of Murcia (A13150102). Before the experimental stage, each animal was housed individually for 5 days, under the optimal vivarium conditions for the detailed specie (temperature, light/darkness cycle, maximum noise, and relative humidity). These conditions are legally established by the EU Directive/63/2010 and by the Spanish Royal Decree 53/2013. During the entire study, animals received ad libitum food and water.

After conditioning, animals were randomly allocated into three groups (n = 5 each) in correspondence to three defined study periods (1, 3 and 5 months) respectively.

2.3. *Scaffold Implantation: Anesthetic and Surgical Method*

All surgical procedures were performed under rigorous aseptic conditions. Animal anesthesia and post-operative analgesia was performed following the previously established and approved protocol described by Ros-Tarraga et al. (2016) [25,26]. Premedication and anesthesia was achieved with atropine sulfate (0.3 mg k^{-1}, im), chlorpromazine hydrochloride (10 mg k^{-1}, im), xylacine (0,25 mg k^{-1}, im) and ketamine hydrochloride (50 mg k^{-1}, im). Animals were treated with a prophylactic single dose of enrofloxacin (mg k^{-1}, im) (Virbac, Barcelona, Spain) to reduce the risk of surgical site infections.

Surgical surface was shaved, washed and sterilized with clorhexidine® (Bohm SA, Madrid, Spain) and povidone iodine 10% (Betadine™; Meda Pharma, Madrid, Spain). Sterile fenestrates adhesive drapes were used for delimitation of surgical area. A 1.5–2 mm long and deep skin incision was made at the proximal anteromedial metaphysis, in parallel to the right tibial shaft axis. Anterior tibial tuberosity was landmarked as surgical reference to minimize incisional variability. After dissection of fascia and periostium, a unicortical ~4.5 mm diameter bone defect was created with a surgical bone drill, coupled to a micromotor at low revolutions and continuous irrigation with saline solution. Bone medullar cavity was not invaded. Surgical defects were debrided and washed with physiological saline solution before being grafted with porous cylindrical implants as it has been shown in Figure 1B. Within each previously defined animal group, the DBM-doped: non-doped scaffold ratio implanted was 3:2. Surgical wound was sutured in anatomical layers with 3-0 Coated Vicryl® and 3-0 Vicryl rapide® (Johnson & Johnson Medical Devices & Diagnostics, New Brunswick, NJ, USA and covered by the application of a thin layer of NovecutanTM plastic dressing spray (Inibsa, Barcelona, Spain). Post-operative analgesia was assessed by the application of subcutaneous mepivacaine (1%) around the surgical wound and buprenorphine (0.3 mg k^{-1}, im, every 12 h for 4 days). After 4 days, individual analgesia was provided in case of pain symptoms, local swelling, or stress. After surgery, limb free movements were permitted. All animals used during the study survived and appeared to be in good health status.

2.4. Euthanasia and Samples Collection

Three different sampling times were considering across the study (1, 3 and 5 months after surgery). For each sampling time, a group of animals (n = 5) was deeply sedated with a single dose of ketamine hydrochloride (50 mg k^{-1}, im) and euthanized by an intracardiac overdose of pentobarbital (Dolethal®, Lab Vetoquinal, Cedex, France). For each animal, right limb tibia was removed, cleaned of soft tissue and fixed in neutral buffered formalin (10%) (Figure 1C). Samples were stored at 4 °C until analyses.

2.5. mCT Imaging Protocol and Descriptive Analysis

The imaging study was performed using the Albira tri-modal preclinical-scanner (Bruker®, Billerica, MA, USA). Fixed scanning parameters were 45 Kv, 0.2 mA, 0.05 mm voxels. From each sample, a set of 1000 axial projections of 0.05 mm thickness was obtained using a digital flat panel detector with 2400 × 2400 pixels and a 70 × 70 mm field of view (FoV). Tibial images were spatially reconstructed by the filtered back projection (FBP) algorithm, and Bone Mineral Density (BMD) in the implanted area was quantified in Hounsfield Unit (HU). With this purpose, a medical image data examiner (AMIDE, UCLA University, LA, USA) and online 3D image analysis software (Volview, mm^3, Inc) were used. Images were pre-descripted by two experienced radiologists (A and B), based on a double-blind visual analysis, evaluating evolution of implanted material along the study (pre-implanted material and 1, 3 and 5 months after surgery). Final report was concluded by consensus among A and B. To homogenize image description, the visual analysis was centered on the following items: (1) loss of homogeneity of implanted material with respect to the pre-implanted one; (2) loss of implant contour sharpness in relation to the peripheric tissue; (3) presence of neoformed bone trabeculations inside the implanted biomaterial; (4) dispersion of biomaterial in the peripheric tissue; (5) neoformed trabeculations between implant surface and adjacent cortical bone. Figure 2 shows the mCT and 3D mCT images evaluated by radiologists.

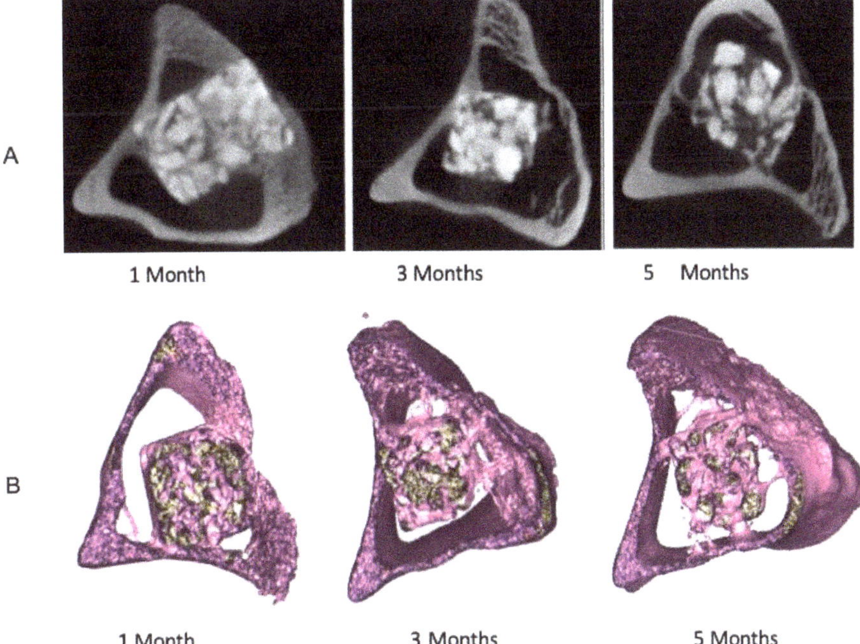

Figure 2. (**A**) Transverse mCT images of implanted area, obtained at different times of study (1, 3 and 5 months); (**B**) 3D mCT image reconstruction of samples obtained at 1, 3 and 5 months of study. Figure show the non-doped scaffold as a representative of both materials.

2.6. Mathematical Processing of mCT Raw Data

Selection and mathematical study of raw data was conducted at the Biostatistics Department of San Antonio Catholic University (Murcia, Spain). With this purpose, three regions of interest (ROIs) were selected by consensus (Figure 3), according to the following criteria: (A) implanted biomaterial (4 × 4 mm^3 cylindrical ROI) to evaluate implant resorption, degradation or integration by bone tissue; (B) implanted biomaterial and peripheral corticomedullar bone (10 × 10 × 10 mm cubical ROI) to assess the dispersion undergone by biomaterial fragments within the peripheral host bone; (C) surgically unaltered cortical bone distanced from the implanted area (3 × 3 mm cylindrical ROI) to determine the spontaneous evolution of unaltered cortical bone, as control measurement and yardstick of analysis. In parallel, two unimplanted porous scaffolds cylinders were analyzed as basal values, being represented as time 0 in the study.

From each ROI and sample, raw data were obtained after voxel selection. According to the basal values obtained, and with the aim of analyzing the interactions between porous scaffold implants and bone tissue, voxels with average values > +1.10 HU were preselected as implanted material from cubical ROI (10 × 10 × 10). In the same way, outranged voxels representing medullar/cortical bone or soft tissue were suppressed. Based on this selection, material dispersion was estimated as follows: A "center of mass" of the whole piece was calculated, determining the mean distance from each preselected voxel to this center. This mathematical study makes it possible to analyze the time evolution of variability in voxel standardized attenuation coefficients (HU), as well as to quantify how much biomaterial had dispersed within the peripheral host bone.

To visually clarify the effect of time on scaffold resorption and dispersion values per each group of implants (DBM-doped and non-doped), trend lines have been added. It must be taken into account that pre-implanted scaffolds were excluded for trend-line adjustment, as resorption and dispersion were not possible.

ROIs selection and raw data were obtained with Albira CT software (Bruker®, Billerica, MA, USA), in conjunction with Volview software (Kitware Inc), and Medical Image Data Analysis software (AMIDE, UCLA University, Los Angeles, CA, USA).

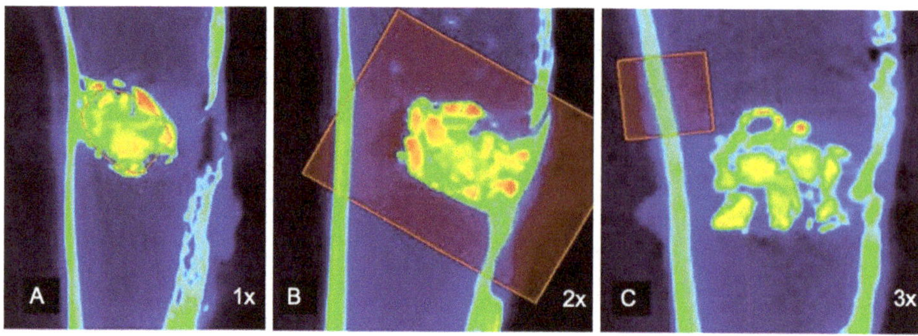

Figure 3. mCT images processed with AMIDE software. Defined Regions of interest (ROIs): (**A**) Implanted biomaterial (Cylindrical 4 × 4 mm), (1×); (**B**) implanted biomaterial and peripheral corticomedullar bone (10 × 10 × 10 mm cubical ROI) (2×). Red squares describe the ROI of cortical area; (**C**) surgically unaltered cortical bone (3 × 3 mm cylindrical ROI) (3×). Red squares describe the cortical area. Different color values correspond to different Hounsfield Unit (HU) ranges: HU > + 1100 (yellow-orange) correspond to different states of the implanted biomaterial. + 0.50–+ 1.00 HU (green) correspond to cortical bone and reabsorbed biomaterial. HU + 0.20–+ 0.40 HU (blue) correspond to connective and medullary bone tissue.

2.7. Histological and Histomorphometric Study

Samples were cross-sectioned perpendicular to the longitudinal axis of tibia with an electric circular saw (Figure 4A). Pieces of 3–4 mm thickness were fixed in neutral buffered formalin (10% formalin in 0.08 M sodium phosphate, pH 7.4) for 48 h and decalcified by the addition of hydrochloride acid (Osteomoll®, Merck Chemical, Darmstadt, Germany). Samples were subsequently dehydrated in alcohol and embedded in paraffin. Histologic and histomorphometric analysis was performed in hematoxylin-eosin stained sections of 4 μm thicknesses. A panoramic histological image (low magnification, ×5.4) of implanted area in each section was obtained (Figure 4B) using a Leica Z6 Apo macroscope (Leica Microsystems, Barcelona, Spain), connected to a Leica CDC500 digital camera and image-capture software (Leica Application Suite). Equipment was provided by the Image Analysis Service of the University of Murcia, Spain (SAI-UMU).

Total implanted area surface (TIS) was calculated, and consecutives microscopic fields were captured under ×50 magnification (Figure 4C). Magnified images were obtained by a bright-field Zeiss Axio Scope A.1 optical microscope (Carl Zeiss, Jenna, Germany), connected to a digital camera (AxioCam IcC3, Carl Zeiss). Equipment was supplied by Pathology Platform of "Instituto Murciano de Investigación Biosanitaria Virgen de la Arrixaca" (IMIB-Arrixaca), Murcia, Spain. As it can be seen in Figure 5, from each 50x field, 4 identifiable components were analyzed: (1) Floury and acidophilic material, identified as "non-absorbed implanted material"; (2) "neoformed bone tissue"; (3) "connective tissue"; (4) floury and basophilic material, located between components 1 and 2, named "unidentifiable material". For components 2, 3 and 4, the surface area was outlined manually using AxioVision rel. 4.8 (Zeiss, Canada) software. To calculate "non-absorbed-implanted material" surface area, the following formula was used:

$$Surface\ area\ (1) = TIS - (2 + 3 + 4) \tag{1}$$

With the aim of an accurate quantification of each component surface extension, the morphometric analysis was repeated on three different levels per sample (inter-level distance 300 μm).

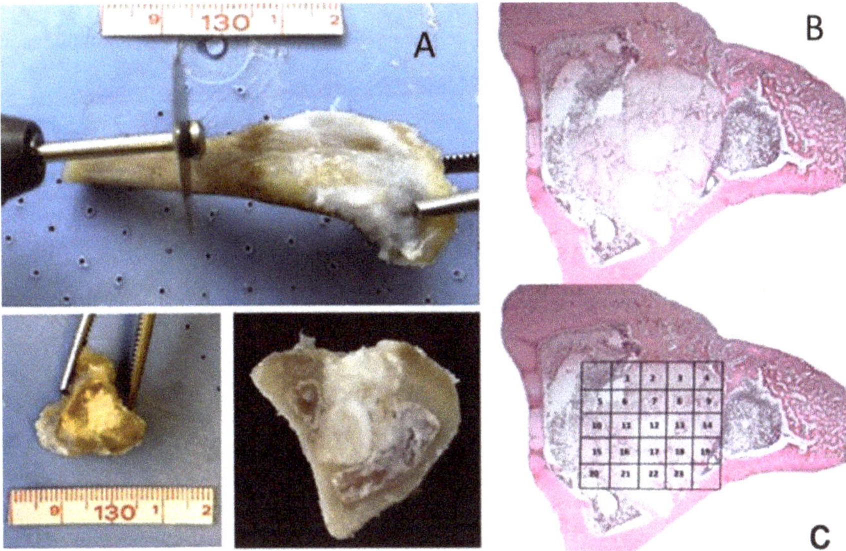

Figure 4. Histological-Histomorphometric procedure. (**A**) Samples including implanted area were obtained perpendicularly to the long axis of tibia, using an electric circular saw; (**B**) Panoramic histological view (×5.4) containing porous material; (**C**) ×50 magnified and consecutives images on a panoramic histological view.

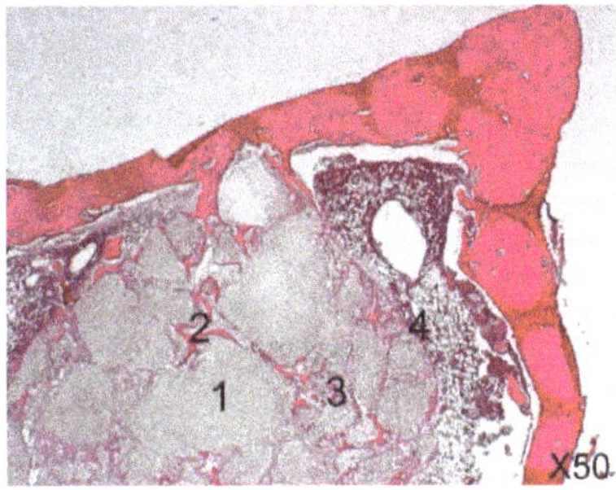

Figure 5. 50× magnified image obtained from the panoramic histological view of the non-doped scaffold as a representative of both materials. Four different components can be identified in the study: (1) non-absorbed implanted material; (2) neoformed bone tissue; (3) connective tissue; (4) special material.

2.8. Statistical Analysis

Statistical analysis was performed using R software v.3.2.3. A previous statistical descriptive analysis was performed to data characterization. Raw data were mathematically processed using a linear regression analysis, which simultaneously included 14,153 points from all the pieces. Linear regression was aimed to detect significant relations ($p < 0.05$) between the average HU of each ROI. As variables, the analysis included: Area location (ROIs), implant composition (DBM-doped or non-doped) and sampling time (non-implanted materials (0), 1, 3 and 5 months). In addition, biomaterial dispersion was analyzed with a to interpretative mathematical study of the biomaterial's

HU values respect the center of mass, establishing its evolution across the study (1, 3 and 5 months) regardless if it is DBM-doped or non-doped

Regarding Histological and Histomorphometric study, a Mann–Whitney non-parametric test was performed. Statistically significant differences (* $p < 0.05$, ** $p < 0.05$) between each previously defined component (1, 2, 3 and 4) in relation to time of sampling (non-implanted material (0), 1, 3 and 5 months) were analyzed.

3. Results

3.1. Descriptive Analysis of mCT Images

mCT images were pre-evaluated by a double-blind interpretation. Final report was issued by consensus based on the previously described procedure. As can be seen on Figure 2A, when mCT images where visually analyzed, both radiologist A and B, concluded that during the 5-month study period, implanted material underwent progressive dispersion into the host bone medulla after matrix disaggregation and disintegration.

According to the final report, when 1-, 3-, and 5-months 3D mCT images (Figure 2B) were visually compared, a progressive disintegration and disorganization of initial implant morphology (from a compact cylindrical block to a fragmented and porous mass) as well as a neoformation of bone tissue were descripted. These findings were more noticeable and earlier detected at the periphery than at the central portion of implanted material. Consensuated descriptional analysis of mCT and 3D mCT images was not able to distinguish differences between the behavior of DMB-doped and no-doped implants across the study period.

Previously, explained findings in descriptive analysis of mCT and 3D mCT (Figure 2) were related to an increment in biomaterial porosity, as well as to the dispersion of implant fragments. This fragmentation seems to be more evident at the scaffold periphery, allowing neoformation of trabecular bone both inside the implanted material and around its surface.

3.2. mCT Raw-Data Analysis

The evolution of scaffold during the study period (pre-implanted material, 1, 3 and 5 months after surgery) was analyzed in the defined cylindrical 4x4mm ROIs, which contained the implanted material (> + 1.10 HU). As can be seen in Figure 6A, a significant relation ($p < 0.05$) was defined in each selected ROI between the average standardized attenuation coefficient (average HU value) and time, accompanied by an increment in voxels HU variability. This relation has been represented by the adjusted trend lines per each group of implants (DBM-doped and non-doped), which suggest a time-dependent decrease of average HU values. To be concrete, pre-implanted scaffolds showed an average attenuation coefficient of 1.13 ± 0.18 HU. Three months after surgery, the attenuation coefficient decreased to average values of 0.99 ± 0.23 HU (DBM-doped scaffolds) and 0.86 ± 0.25 HU (non-doped scaffolds). Five-month attenuation coefficients reached minimum values of 0.86 ± 0.32 HU (DBM-doped scaffolds) and 0.66 ± 0.33 HU (non-doped scaffolds). As can be deduced from these results, DBM-doped scaffolds showed a more gradual decrease of average HU values than non-doped implants, resulting in significant differences ($p < 0.05$) between both groups average values.

Biomaterial dispersion in peripheral tissue was analyzed selecting cubical ROIs ($10 \times 10 \times 10$ mm^3) and a > + 1.10 HU window. As can be seen in Figure 6B, the variation of global distance (mm) of selected voxels with respect to the center of mass of implanted material, increased at a rate of ~ 0.16 mm per month, from the beginning (1 month) to the end (5 months) of post-surgical period analyzed. The dispersion value increment between different stages of study were statistically significant ($p < 0.05$); however, no differences were found between DBM-doped and no-doped implants.

As control measurement, the average standardized attenuation coefficient in surgically unaltered cortical bone (0.50–1.00 HU window) was analyzed in cylindrical ROI (3×3 mm^3). According to the results obtained, average unaltered cortical HU values remained stable across the study (0.83 ± 0.12 HU), with no significant differences between different sampling times (1, 3 and 5 months after surgery).

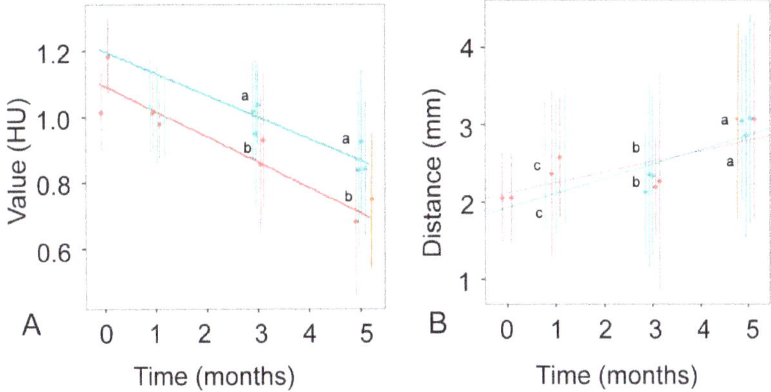

Figure 6. Evolution of implanted material after 1, 3, and 5 months post-implantation in each animal group ($n = 5$). Pre-implanted material results are considered as time 0. Doped (Activagen®-treated) and no-doped material have been represented as blue and red dots, respectively. (**A**) For evaluation of implant resorption, results have been presented as the average HU ± S.D. in each preselected cylindrical ROI (4 × 4 mm^3). Different letters a, b denote significant differences between type of material for each time of study. (**B**) Per each cubical ROI (10 × 10 × 10 mm^3), implanted material dispersion in the peripheral tissue has been presented as the center of mass per each implant (dot) and the global variation of distance (mm) of the preselected voxels (>1.10 HU) from this center of mass (bars). A > 1.10 HU window was set to identify voxels corresponding to biomaterial. Different letters a, b, c denote significant differences between time).

3.3. Histological and Histomorphometrical Results

According to Figure 7A, there was a good visual correspondence between mCT images and histological preparations. Implanted scaffolds were degraded and resorbed during the time of study, being transformed into fragments which were smaller at the periphery and larger at the central portion of the implant. When the sites between remaining scaffold fragments were histologically analyzed (Figure 7B,C), it consisted of osteogenic tissue (fibroblasts, osteoclasts and osteoblast, surrounded by extracellular matrix) with signs of active angiogenesis.

Histological image of implant was divided in consecutives ×50 microscopic fields for histomorphometric analysis (Figure 4C). The surface area of each histologic component considered in the study [(A) non-absorbed implanted material; (B) neoformed bone tissue; (C) connective tissue; (D) unidentifiable material] was calculated. In Figure 8, the evolution of these components surfaces with time of study (1, 3 and 5 months) have been shown. A progressive and significant increase in bone tissue (Figure 8B) was observed from 1 to 5 month after surgery. In this way, bone tissue significantly increased ($p < 0.05$) from 1.27 ± 0.30 mm^2 (non-doped) and 1.09 ± 0.32 mm^2 (DBM-doped), to 2.12 ± 0.21 mm^2 (non-doped) and 2.15 ± 0.35 mm^2 (DBM-doped) 3 months post-surgery. This increment become more evident 5 months post-surgery ($p < 0.005$), reaching final values of 3.91 ± 0.39 mm^2 (non-doped) and 3.43 ± 0.27 mm^2 (DBM-doped).

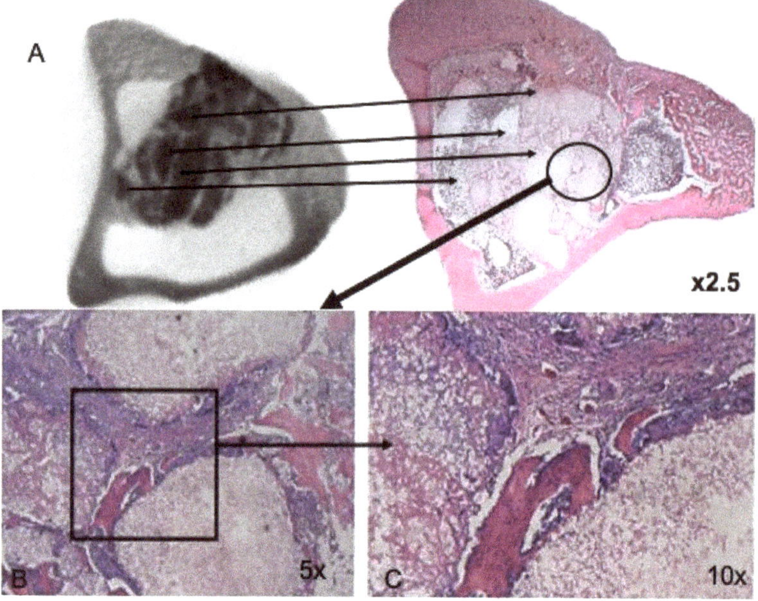

Figure 7. (**A**) Visual correlation between histological (hematoxylin and eosin stain) and axial mCT images. (**B**,**C**) Detailed view (×5 and ×10 magnification) of area contained between scaffolds fragments. Corresponding to the non-doped scaffold as a representative of both materials.

In parallel to bone tissue increment, a significant progressive reduction in surface of implanted material (Figure 8A) and basophilic component surrounding neoformed bone tissue (unidentifiable material) (Figure 8D) can be described. Regarding implanted material, its surface decreased from 27.21 ± 5.28 mm^2 (non-doped) and 26.92 ± 4.19 mm^2 (DBM-doped), to 15.88 ± 5.01 mm^2 (non-doped) and 16.52 ± 3.28 mm^2 (DBM-doped) 5 months post-surgery. Similar behavior was described for unidentifiable material surface, decreasing from 1.75 ± 0.90 mm^2 (non-doped) and 1.64 ± 0.46 mm^2 DBM-doped) at 1 month, to a final value of 0.30 ± 0.29 mm^2 (non-doped) and 0.36 ± 0.23 mm^2 (DBM-doped) 5 months post-surgery.

When the post-surgery evolution of surface occupied by connective tissue was analyzed (Figure 8C), significant differences were found between 1 month (1.10 ± 0.75 and 1.22 ± 0.41 mm^2, non-doped and DBM-doped respectively) and 3 months (1.63 ± 0.92 and 1.78 ± 0.71 mm^2, non-doped and DBM-doped material respectively). However, at 5 months after surgery, connective tissue decreased to 0.43 ± 0.29 mm^2 (non-doped) and 0.68 ± 0.26 mm^2 (DBM-doped). Five-month results, were statistically significant ($p < 0.005$) only when compared to DBM-doped implanted materials at 1 and 3 months after surgery.

After the analysis of the effect of scaffold composition (non-doped and DBM-doped material), no statistical differences were found for each time considered in the study (1, 3 and 5 months after surgery). Despite this fact, when average values for each material after 5 months were compared, DBM-doped scaffolds resulted in an average lower reduction of connective tissue (Figure 8C) and a lower bone tissue formation (Figure 8B), than the average values obtained for non-doped implant. Differences in average values between DBM-doped and non-doped materials, were less noticeable for the evolution of surface area of residual implanted material (Figure 8A) and unidentifiable material (Figure 8D), than for the evolution of connective tissue and bone tissue formation.

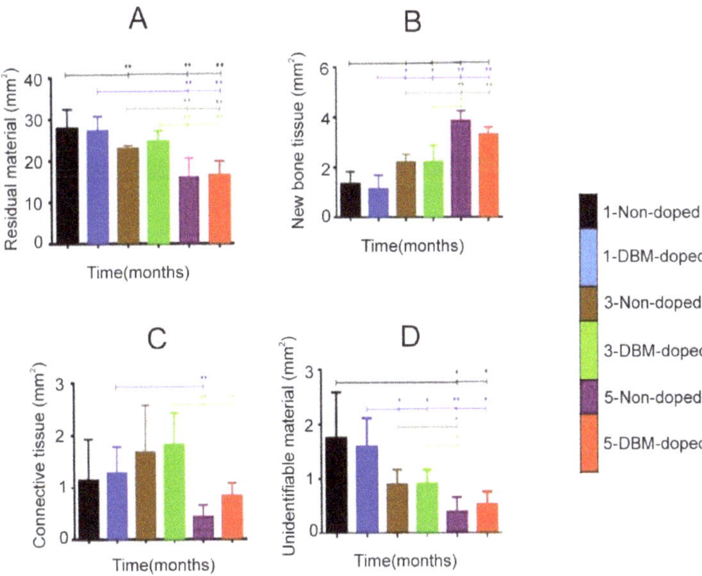

Figure 8. Surface area (mm^2) evolution of different histological components. (**A**) residual biomaterial, (**B**) new bone tissue, (**C**) connective tissue and (**D**) unidentifiable material in sliced histological samples]. Results have been presented as means ± S.D. Horizontals bars represent the Mann–Whitney–Wilcoxon test for two samples comparison (* $p < 0.05$ and ** $p < 0.005$).

4. Discussion

As has been widely described, bone repair process consists of (1) the recruitment of mesenchymal cells, which differentiates into fibroblast and osteogenic cells, (2) extracellular chondroid matrix formation, and (3) mineralization/chondroid matrix resorption [27]. Current bone tissue engineering includes the implantation of porous calcium phosphate and silicate scaffolds. Their presence induces a response similar to the one described for bone remodeling [28–32]. It has been described that when porous scaffolds are implanted, they are progressively resorbed and substituted by neoformed bone tissue. These process follows a concentric evolution, from the periphery to the center of the implanted material [33–35]. The results of this work agree with this process. After implantation, scaffolds were progressively degraded from 1 to 5 months after surgery. This degradation commenced at the periphery of the implanted material and advanced to its nucleus. Consensuated descriptive analysis of 3D mCT was able to describe this process as a progressive disintegration and disorganization of scaffold initial morphology, accompanied by the neoformation of bone tissue. However, descriptive analysis was not able neither, to detect the evolution of the connective matrix, nor quantify the time evolution of scaffold degradation and dispersion/bone neoformation. To clarify these issues, it was necessary to resort to histomorphometric and mCT raw-data analysis. The need to compliment the mCT results with other techniques for in vivo time scaffold evolution can be noticed in the previously published bibliography [30,36] including mathematical image analysis, histological macro and microscopic analysis, or design of specific algorithm for mCT data analysis.

As it has been explained, raw-data analysis showed that scaffold dispersion took place at a rate of ~ 0.16 mm per month. This material dispersion reflects the implanted material degradation. Considering a total scaffold surface area of 116.63 mm^2, and according to histomorphometric results, implant surface showed an approximate time-dependent degradation of 12-24% after 5 months post-implantation. These results can be partially compared to those showed by Sweedy et al. (2017) [22], who defined a scaffold resorption ratio of 22.16 ± 3.3%–31.66 ± 6.8% for a 6 weeks study period using Swiss Alpine sheep. However, it must be taken into account that this resorption ratio not only depends on animal species, but also is affected by scaffold porous size, implant composition or scaffold total size [37–39]. Furthermore, Bohner et al. (2017) [21] described that scaffold resorption and bone/mineralized tissue formation is not a uniform process, with a rapid increase in the first 4–10 weeks followed by a slow decrease after 10–12 weeks post-implantation. Based in the previously exposed results, scaffold degradation was uniform during the 5-month post-surgery period. By contrast, bone tissue formation and extracellular matrix resorption of the physical characteristics of scaffolds (e.g., micro- and macropore diameter) were not set during this study, and the differences between both scaffolds in resorption/bone repair process evolution cannot be really discussed. Bohner et al. (2017) [21] concluded in their work that mineralized matrix and ulterior bone formation needed an interconnected porous net, with a pore size larger than 1–10 μm. This would mean that a previous scaffold degradation process is needed to provide a suitable structure for bone repair, explaining the progressive scaffold degradation and the initial slow extracellular matrix resorption and bone formation previously described.

Regarding scaffold composition, the treatment with DBM-doped did not result in significant differences between doped and non-doped implants. Only raw-data mathematical analysis was able to detect significant differences in scaffold HU average values across the study between both types of materials. In fact, doped materials resulted in a more gradual decrease of average HU than no-doped implants. These results could be due to a higher phagocytic and osteoclastic resorption when non-doped scaffolds were implanted. However, no significant differences were described when implant dispersion, extracellular matrix resorption and bone tissue formation were analyzed. DBM-doped is composed of lyophilized granules of collagen type I [12]. After implantation to bone defects, biomaterial is degraded by osteoclasts, leading to the release of protein factors which induces mesenchymal cells differentiation into osteoblasts [30,31,38,40]. Shalash et al. (2013) [33], published a 6-month study based on the comparison between the use of TCP alone, and TCP plus DBM, for maxillary alveolar ridges deficiencies regeneration before implants placement. According to this study, after 6 months, TCP plus DBM resulted in a higher bone formation and TCP resorption than TCP alone. As it has been previously exposed, implantation of non-doped biomaterial, resulted in a higher resorption of scaffolds with no differences in new bone formation. Although the reason for these results remains unclear, a possible explanation could be the early stimulation of osteoblasts differentiation when DBM is added, controlling resorptive activity of osteoclasts [41]. It must be taken into account that the study was performed over 5 months, and synthesis of new bone tissue seemed to be more active at the end of this period. To obtain significant differences in new bone formation between both materials (doped and non-doped), a longer study would be needed.

Beyond these results, it is important to point out that medical invasive techniques are being substituted by minimally or non-invasive alternatives. Recent advances in CT technology are contributing to this purpose. Clear examples can be found in CT angiography, which is a non-invasive alternative to classical coronary angiography [35,36]. Applied to biomedical experimentation on animal models, improvement of imaging analysis techniques can replace invasive histological studies. This would lead to the refinement of experiments, due to the reduction of the number of animals needed, and to the continuous data collection without the need for animal sacrifice at intermediate experimental points [37]. Regarding in vivo assessment of bone quality and regeneration, preclinical research requires the combination of image and histomorphometric analysis [37–40]. Against visual image analysis and volumetric estimation of average standardized X-ray attenuation coefficients, histology provides a direct method to assess the bone regeneration process at a macroscopic and microscopic level, without resolution or standardization limitations. However, histomorphometric analysis requires to foreseeing the sacrifice of different groups of animals during the experiment, preventing from a continuous study and increasing both intragroup variability and number of animals needed [40–42]. Mathematical analysis of CT raw data could help to refine preclinical assays in bone regeneration studies. Their mathematical analysis could improve research based on image analysis. Budán et al. (2018) [43] performed an experiment with rats to develop a new method for evaluation of bone regeneration. With this purpose, CT and single photon emission computed tomography (SPECT) with marked methyl diphosphonate (MDP) as tracer were combined. Based on the cubic voxel reconstruction and calculation of summarized absorbance of VOI after normalization of bone total X-ray attenuation, quantification of bone regeneration was performed through the bone opacity changes, as well as marked MDP activity. Despite normalization, authors declared that as raw data were visually compared, not enough information was provided for proper quantitative evaluation and significant statistical analysis, especially during the early period of the experiment. The mathematical analysis of raw data could help to could help avoid these limitations in the methodology.

5. Conclusions

Mathematical analysis of CT raw data is often used separately by researchers, with no connection to radiological studies. Nevertheless, an adequate exploration of raw data seems to provide an objective analysis of radiological image. The present study shows the necessity of combining both radiological imaging studies and mathematical analysis of CT raw data to perform an adequate in vivo analysis of implanted bone scaffold and its evolution after surgery. Furthermore, as the results obtained seem to be similar to the anatomopathological ones, mathematical analysis of CT raw data would allow the conducting of long-term duration in vivo studies, without the need for animal sacrifice and the subsequent reduction in variability.

The combination of imaging and mathematical analysis could be extended to other areas in radiology practice, allowing clinicians to obtain a more accurate and objective image description, as well as to quantify the evolution of lesions or implants, without having to resort to aggressive techniques such as biopsy. In this way, future studies would be needed to assess these applications, and standardize the mathematical analysis of raw data applied to the radiological practice.

Author Contributions: Conceptualization: A.P.-A., C.A.G.-B.; methodology, S.S.-S., L.M.-O., C.M.M.-C., F.M.-M.; software, J.L.C.-G., J.J.P.d.A., N.G.-C. and J.M.A.; validation, J.J.P.d.A., N.G.-C., P.N.D.A.; formal analysis, J.J.P.d.A., C.A.G.-B.; investigation, A.P.-A., C.A.G.-B., S.S.-S.; resources, C.M.M.-C., F.M.-M., data curation, N.G.-C., P.N.D.A., A.P.-A.; writing—original draft preparation, A.P.-A., J.J.P.d.A., N.G.-C., P.N.D.A.; writing—review and editing, F.M.-M., J.L.C.-G., L.M.O.; visualization, F.M.-M., J.L.C.-G., N.G.-C., P.N.D.A.; supervision, J.L.C.-G., P.N.D.A. and J.M.A.; project administration, P.N.D.A., A.P.-A., C.A.G.-B., S.S.-S.; funding acquisition, P.N.D.A.

Funding: This research was funded by Spanish Ministry of Economy and Competitiveness (MINECO) contract grant number: MAT2013-48426-C2-2-R.

Acknowledgments: We want to thank to the department of Biostatistical of UCAM for its invaluable support in this work.

Conflicts of Interest: The authors declare no conflict of interest.

References

1. Lehmann, T.M.; Meinzer, H.P.; Tolxdorff, T. Advances in biomedical image analysis ast present and future challenges. *Methods Inf. Med.* **2004**, *43*, 308–314.
2. Lehmann, T.M.; Handels, H.; Maier-Hein ne Fritzsche, K.H.; Mersmann, S.; Palm, C.; Tolxdorff, T.; Wagenknecht, G.; Wittenberg, T. Viewpoints on medical image processing: From science to application. *Curr. Med. Imaging Rev.* **2013**, *9*, 79–88. [CrossRef]
3. Bouxsein, M.L.; Boyd, S.K.; Christiansen, B.A.; Guldberg, R.E.; Jepsen, K.J.; Müller, R. Guidelines for assessment of bone microstructure in rodents using microcomputed tomography. *J. Bone Miner. Res.* **2010**, *25*, 1468–1486. [CrossRef]
4. Höhne, K.H.; Bomans, M.; Pommert, A.; Riemer, M.; Schiers, C.; Tiede, U. 3D visualization of tomographic volume data using the generalized voxel model. *Vis. Comput.* **1990**, *6*, 28–36. [CrossRef]
5. Beister, M.; Kolditz, D.; Kalender, W.A. Iterative reconstruction methods in X-ray CT. *Phys. Med.* **2012**, *28*, 94–108. [CrossRef]
6. Garland, L.H. On the scientific evaluation of diagnostic procedures: Presidential address thirty-fourth annual meeting of the Radiological Society of North America. *Radiology* **1949**, *52*, 309–328. [CrossRef]
7. Berlin, L. Radiologic errors, past, present and future. *Diagnosis* **2014**, *1*, 79–84. [CrossRef]

8. Bruno, M.A.; Walker, E.A.; Abujudeh, H.H. Understanding and confronting our mistakes: The epidemiology of error in radiology and strategies for error reduction. *RadioGraphics* **2015**, *35*, 1668–1676. [CrossRef]
9. Waite, S.; Scott, J.; Gale, B.; Fuchs, T.; Kolla, S.; Reede, D. Interpretive error in radiology. *Am. J. Roentgenol.* **2017**, *208*, 739–749. [CrossRef]
10. Castellino, R.A. Computer aided detection (CAD): An overview. *Cancer Imaging* **2005**, *5*, 17–19. [CrossRef]
11. Doi, K. Computer-aided diagnosis in medical imaging: Historical review, current status and future potential. *Comput. Med. Imaging Graph.* **2007**, *31*, 198–211. [CrossRef]
12. Parrilla-Almansa, A.; García-Carrillo, N.; Ros-Tárraga, P.; Martínez, C.M.; Martínez-Martínez, F.; Meseguer-Olmo, L.; de Aza, P.N. Demineralized bone matrix coating Si-Ca-P ceramic does not improve the osseointegration of the scaffold. *Materials* **2018**, *11*, 1580. [CrossRef]
13. Martinez, I.M.; Velasquez, P.A.; de Aza, P.N. Synthesis and stability of α-tricalcium phosphate doped with dicalcium silicate in the system $Ca_3(PO_4)_2$–Ca_2SiO_4. *Mater. Charact.* **2010**, *61*, 761–767. [CrossRef]
14. Kim, S.Y.; Kim, Y.K.; Park, Y.H.; Park, J.C.; Ku, J.K.; Um, I.W.; Kim, J.Y. Evaluation of the healing potential of demineralized dentin matrix fixed with recombinant human bone. Morphogenetic protein-2 in bone grafts. *Materials* **2017**, *10*, 1049. [CrossRef]
15. Schmidt, R.; Muller, L.; Kress, A.; Hirschfelder, H.; Aplas, A.; Pitto, R.P. A computed tomography assessment of femoral and acetabular bone changes after total hip arthroplasty. *Int. Orthop.* **2002**, *26*, 299–302. [CrossRef]
16. Mate-Sanchez de Val, J.E.; Calvo-Guirado, J.L.; Delgado-Ruiz, R.A.; Ramirez-Fernandez, M.P.; Negri, B.; Abboud, M.; Martinez, I.M.; de Aza, P.N. Physical properties, mechanical behavior, and electron microscopy study of a new α-TCP block graft with silicon in an animal model. *J. Biomed. Mater. Res. A* **2012**, *100*, 3446–3454. [CrossRef]
17. Calvo-Guirado, J.L.; Ramirez-Fernandez, M.P.; Delgado-Ruiz, R.A.; Mate-Sanchez de Val, J.E.; Velasquez, P.; de Aza, P.N. Influence of biphasic β-TCP with and without the use of collagen membranes on bone healing of surgically critical size defects. A radiological, histological and histomorphometric study. *Clin. Oral Implant Res.* **2014**, *25*, 1228–1238. [CrossRef]
18. Ros-Tarraga, P.; Mazon, P.; Rodriguez, M.A.; Meseguer-Olmo, L.; de Aza, P.N. Novel resorbable and osteoconductive calcium silicophosphate scaffold induced bone formation. *Materials* **2016**, *9*, 785. [CrossRef]
19. Marsell, R.; Einhorn, T.A. The biology of fracture healing. *Injury* **2011**, *42*, 551–555. [CrossRef]
20. Ramirez-Fernandez, M.P.; Mazon, P.; Gehrke, S.A.; Calvo-Guirado, J.L.; de Aza, P.N. Comparison of two xenograft materials used in sinus lift procedures: Material characterization and in vivo behavior. *Materials* **2017**, *10*, 623. [CrossRef]
21. Bohner, M.; Baroud, G.; Bernstein, A.; Dobelin, N.; Galea, L.; Hesse, B.; Heuberger, R.; Meille, S.; Miche, P.; von Rechenberg, B.; et al. Characterization and distribution of mechanically competent mineralized tissue in micropores of β-tricalcium phosphate bone substitutes. *Mater. Today* **2017**, *20*, 106–115. [CrossRef]
22. Sweedy, A.; Bohner, M.; Baroud, G. Multimodal analysis of in vivo resorbable CaP bone substitutes by combining histology, SEM, and microcomputed tomography data. *J. Biomed. Mater. Res. Part B Appl. Biomater.* **2017**, *106*, 1567–1577. [CrossRef]
23. Mate-Sanchez de Val, J.E.; Mazón, P.; Piattelli, A.; Calvo-Guirado, J.L.; Bueno, J.M.; de Aza, P.N. Comparison among the physical properties of calcium phosphate-based bone substitutes of natural or synthetic origin. *Int. J. Appl. Ceram. Technol.* **2018**, *15*, 930–937. [CrossRef]
24. Calvo-Guirado, J.L.; Ballester-Montilla, A.M.; de Aza, P.N.; Fernández-Domínguez, M.; Gehrke, S.A.; Cegarra-Del Pino, P.; Mahesh, L.; Pelegrine, A.A.; Aragoneses, J.M.; Maté-Sánchez de Val, J.E. Particulate extracted human teeth characterization by SEM-EDX evaluation as a biomaterial for socket preservatio: An in vitro study. *Materials* **2019**, *12*, 380. [CrossRef]
25. Mate Sanchez de Val, J.E.; Calvo Guirado, J.L.; Delgado Ruiz, R.A.; Gomez Moreno, G.; Ramírez Fernández, M.P.; Romanos, G.E. Bone neo-formation and mineral degradation of 4Bone.®Part I: Material characterization and SEM study in critical size defects in rabbits. *Clin. Oral. Implants Res.* **2015**, *26*, 116–1169.
26. Barone, A.; Toti, P.; Quaranta, A.; Alfonsi, F.; Cucchi, A.; Calvo-Guirado, J.L.; Negri, B.; Di Felice, R.; Covani, U. Volumetric analysis of remodelling pattern after ridge preservation comparing use of two types of xenografts. A multicentre randomized clinical trial. *Clin Oral Implants Res.* **2016**, *27*, e105–e115. [CrossRef]

27. Dozza, B.; Lesci, I.G.; Duchi, S.; Della Bella, E.; Martini, L.; Salamanna, F.; Falconi, M.; Cinotti, S.; Fini, M.; Lucarelli, E.; et al. When size matters: Differences in demineralized bone matrix particles affect collagen structure, mesenchymal stem cell behavior, and osteogenic potential. *J. Biomed. Mater. Res. Part A* **2017**, *105*, 1019–1033. [CrossRef]
28. de Aza, P.N.; Rodríguez, M.A.; Gehrke, S.A.; Mate-Sanchez de Val, J.E.; Calvo-Guirado, J.L. A Si-αTCP scaffold for biomedical applications: An experimental study using the rabbit tibia model. *Appl. Sci.* **2017**, *7*, 706. [CrossRef]
29. Velasquez, P.; Luklinska, Z.B.; Meseguer-Olmo, L.; Mate-Sanchez de Val, J.E.; Delgado-Ruiz, R.A.; Calvo-Guirado, J.L.; Ramirez-Fernandez, M.P.; de Aza, P.N. αTCP ceramic doped with dicalcium silicate for bone regeneration applications prepared by powder metallurgy method: In vitro and in vivo studies. *J. Biomed. Mater. Res. Part A* **2013**, *101*, 1943–1954. [CrossRef]
30. Lugo, G.J.; Mazón, P.; de Aza, P.N. Phase transitions in single phase Si-Ca-P-based ceramic under thermal treatment. *J. Eur. Ceram. Soc.* **2015**, *35*, 3693–3700. [CrossRef]
31. Lugo, G.J.; Mazón, P.; de Aza, P.N. Material processing of a new calcium silicophosphate ceramic. *Ceram. Int.* **2016**, *42*, 673–680. [CrossRef]
32. Rubio, V.; de la Casa-Lillo, M.A.; de Aza, S.; de Aza, P.N. The system $Ca_3(PO_4)_2$–Ca_2SiO_4: The sub-system Ca_2SiO_4-7CaO$P_2O_5$2SiO_2. *J. Am. Ceram. Soc.* **2011**, *94*, 4459–4462. [CrossRef]
33. Shalash, M.A.; Rahman, H.A.; Azim, A.A.; Neemat, A.H.; Hawary, H.E.; Nasry, S.A. Evaluation of horizontal ridge augmentation using beta tricalcium phosphate and demineralized bone matrix: A comparative study. *J. Clin. Exp. Dent.* **2013**, *5*, e253–e259. [CrossRef]
34. Andersen, T.L.; Del Carmen Ovejero, M.; Kirkegaard, T.; Lenhard, T.; Foged, N.T.; Delaissé, J.M. A scrutiny of matrix metalloproteinases in osteoclasts: Evidence for heterogeneity and for the presence of MMPs synthesized by other cells. *Bone* **2004**, *35*, 1107–1119. [CrossRef]
35. Benedek, I.; Chitu, M.; Kovacs, I.; Balazs, B.; Benedek, T. Incremental value of preprocedural coronary computed tomographic angiography to classical coronary angiography for prediction of PCI complexity in left main stenosis. *World J. Cardiovasc. Dis.* **2013**, *3*, 573–580. [CrossRef]
36. Chow, B.J.; Hoffmann, U.; Nieman, K. Computed tomographic coronary angiography: An alternative to invasive coronary angiography. *Can. J. Cardiol.* **2005**, *21*, 933–940.
37. Beckmann, N.; Ledermann, B. Noninvasive small rodent imaging: Significance for the 3R principles. In *Small Animal Imaging*; Kiessling, F., Pichler, B., Hauff, P., Eds.; Springer: Berlin, Germany, 2017.
38. Oryan, A.; Eslaminejad, M.B.; Kamali, A.; Hosseini, S.; Moshiri, A.; Baharvand, H. Mesenchymal stem cells seeded onto tissue-engineered osteoinductive scaffolds enhance the healing process of critical-sized radial bone defects in rat. *Cell Tissue Res.* **2018**, *374*, 63–81. [CrossRef]
39. González-Gil, A.B.; Lamo-Espinosa, J.M.; Muiños-López, E.; Ripalda-Cemboráin, P.; Abizanda, G.; Valdés-Fernández, J.; López-Martínez, T.; Flandes-Iparraguirre, M.; Andreu, I.; Elizalde, M.R.; et al. Periosteum-derived mesenchymal progenitor cells in engineered implants promote fracture healing in a critical-size defect rat model. *J. Tissue Eng. Regen. Med.* **2019**. [CrossRef]
40. Develos Godoy, D.J.; Banlunara, W.; Jaroenporn, S.; Sangvanich, P.; Thunyakitpisal, P. Collagen and mPCL-TCP scaffolds induced differential bone regeneration in ovary-intact and ovariectomized rats. *Bio-Med. Mater. Eng.* **2018**, *29*, 389–399. [CrossRef]
41. Francois, E.L.; Yaszemski, M.J. Preclinical bone repair models in regenerative medicine. *Princ. Regen. Med.* **2019**, *43*, 761–767.
42. Ruehe, B.; Niehues, S.; Heberer, S.; Nelson, K. Miniature pigs as an animal model for implant research: Bone regeneration in critical-size defects. *Oral Surg. Oral Med. Oral Pathol. Oral Radiol. Endod.* **2009**, *108*, 699–706. [CrossRef]
43. Budán, F.; Szigeti, K.; Weszl, M.; Horváth, I.; Balogh, E.; Kanaan, R.; Berényi, K.; Lacza, Z.; Máthé, D.; Gyöngyi, Z. Novel radiomics evaluation of bone formation utilizing multimodal (SPECT/X-ray CT) in vivo imaging. *PLoS ONE* **2018**, *13*, e0204423. [CrossRef]

Article

Vertical Bone Construction with Bone Marrow-Derived and Adipose Tissue-Derived Stem Cells

Thaiz Carrera-Arrabal [1], José Luis Calvo-Guirado [2], Fabricio Passador-Santos [1], Carlos Eduardo Sorgi da Costa [1], Frank Róger Teles Costa [1], Antonio Carlos Aloise [1], Marcelo Henrique Napimoga [1], Juan Manuel Aragoneses [3] and André Antonio Pelegrine [1,*]

1 Faculdade São Leopoldo Mandic, Instituto de Pesquisas São Leopoldo Mandic, Campinas 13045-755, Brazil; thaizarrabal@gmail.com (T.C.-A.); fabricio.passador-santos@slmandic.edu.br (F.P.-S.); du_studio_oral@yahoo.com.br (C.E.S.d.C.); frankrogertc@gmail.com (F.R.T.C.); aca.orto@uol.com.br (A.C.A.); marcelo.napimoga@slmandic.edu.br (M.H.N.)

2 Department of Oral and Implant Surgery. Faculty of Health Sciences, Universidad Católica San Antonio de Murcia (UCAM), 30002 Murcia, Spain; jlcalvo@ucam.edu

3 Department of Dental Research in Universidad Federico Henríquez y Carvajal (UFHEC), Santo Domingo 10107, Dominican Republic; jmaragoneses@gmail.com

* Correspondence: andre.pelegrine@slmandic.edu.br; Tel.: +55-19-981737532

Received: 4 December 2018; Accepted: 29 December 2018; Published: 8 January 2019

Abstract: The purpose of this study was to conduct a histomorphometric analysis of bone marrow-derived and adipose tissue-derived stem cells, associated with a xenograft block, in vertical bone constructions in rabbit calvaria. Ten rabbits received two xenograft blocks on the calvaria, after decortication of the parietal bone. The blocks were fixed with titanium screws. The blocks were combined with the bone marrow-derived mesenchymal stem cells in the bone marrow stem cell (BMSC) group (right side of the calvaria) or with the adipose tissue-derived mesenchymal stem cells in the adipose tissue stem cell (ATSC) group (left side of the calvaria). After 8 weeks, the animals were sacrificed and their parietal bones were fixed in 10% formalin for the histomorphometric analysis. The following parameters were evaluated—newly formed bone (NFB), xenogeneic residual particles (XRP), and non-mineralized tissue (NMT). The histomorphometric analysis revealed 11.9 $\pm$ 7.5% and 7.6 $\pm$ 5.6% for NFB, 22.14 $\pm$ 8.5% and 21.6 $\pm$ 8.5% for XRP, and 65.8 $\pm$ 10.4% and 70.8 $\pm$ 7.4% for NMT in groups BMSC and ATSC, respectively, with statistically significant differences in the NFB and the NMT between the groups, but no differences in the XRP. Therefore, it can be concluded that the bone marrow-derived stem cells seem to have more potential for the bone formation than do the adipose tissue-derived stem cells when used in combination with the xenogenous blocks in the vertical bone construction.

Keywords: bone marrow cells; grafts; adipose tissue-derived stem cells

---

1. Introduction

Large bone constructions represent a challenge for the implant therapy team. In these situations, a bone graft is commonly considered the biological gold standard, once it has osteogenic, osteoinductive, and osteoconductive potentials [1]. However, autografts have a few disadvantages, such as prolonged surgical time and the need for an additional surgery to harvest tissue from the patient at the donor site, which can result in higher morbidity [2,3]. Thus, the study of different types of cell therapy is justified as they present minimal donor site morbidity and a lower risk of autoimmune rejection and disease transmission [4].

Concerning the usage of cell therapy in dentistry, there are many studies on the potential of stem cells for the regeneration of some tissues, such as periodontal tissues [5,6], bone [7,8], and the dentin–pulp complex [9,10]. Most of these studies focused on the use of stem cells from different sources (e.g., bone marrow, adipose tissue, periodontal ligament, and pulp). However, there are a limited number of studies comparing the results of tissue regeneration with stem cells from different sources.

Most bone substitute biomaterials have only osteoconductive potential due to the lack of proteins and of a cellular component [11]. However, the adjunctive use of stem cells with the possibility of osteoblastic differentiation could theoretically result in a composite graft (i.e., stem cell scaffold construct) with osteoconductive, osteoinductive, and osteogenic potential, as demonstrated by Victorelli et al. [12]. Autologous stem cells are adult stem cells that are considered undifferentiated cells found in specialized postnatal tissues and organs. Autologous mesenchymal stem cells have the capacity to differentiate into specialized cells of at least one mesenchymal lineage such as bone, cartilage, fat, or muscle [13], and they can be found in some types of postnatal tissues, such as bone marrow [14], adipose tissue [15], dental tissue [16], and gingival tissue [17].

Therefore, studies comparing the capacity of mesenchymal stem cells from different sources are of major importance in bone tissue engineering, especially in critical situations such as vertical bone constructions. In this study, as in a previous study by our group [7], we compared two of the most frequent types of tissue used in cell therapy—adipose tissue and bone marrow. However, as the delivery vehicle is important to the performance of mesenchymal stem cells [18,19], in the present study, we used a scaffold in block form, which is a common clinical strategy when treating bone defects that require appositional reconstructions.

2. Materials and Methods

This study was analyzed and approved by the Research Ethics Committee of the São Leopoldo Mandic Dental School, Campinas, SP, Brazil (process 0191/14).

2.1. Bone Marrow Harvest by Aspiration

Autologous bone marrow was obtained by aspiration after anesthesia in all 10 animals. Anesthesia was induced with ketamine (40 mg/kg), midazolam (2 mg/kg), and fentanyl citrate (0.8 µg/kg), and maintained with isoflurane/nitrous oxide (1:1.5%) and oxygen (2/3:1/3) with a pediatric facemask. In addition, a local anesthesia was provided via 1 mL of 2% lidocaine HCl and epinephrine 1:100,000 diluted in 1 mL of physiological saline solution.

Two-milliliters of bone marrow aspirates were obtained from each tibia of the ten rabbits using disposable 40 × 10 needles (1.10 mm × 38 mm) and 20-mL disposable syringes previously heparinized to prevent blood clotting.

2.2. Culture of Adult Bone Marrow-Derived Mesenchymal Stem Cells

The procedure to obtain the BMSCs followed the standard guidelines and was completely described in a previous publication of our group (Coelho de Faria et al.) [7].

The BMMSCs, after culture, were detached and resuspended in the culture medium and subsequently mixed with the xenograft in the BMSC group.

2.3. Lipectomy for Adipose Tissue Isolation

The adipose tissue was obtained from the back of all the 10 animals following the same general anesthesia protocol previously described by our group (Coelho de Faria et al.) [7].

The collected material of all the animals was immediately taken to the cell culture laboratory for tissue processing.

2.4. Culture of Adult Adipose Tissue-Derived Mesenchymal Stem Cells

The procedure to obtain the ATSCs followed the standard guidelines and was completely described in a previous publication of our group (Coelho de Faria et al.) [7].

The ATSCs, after culture, were detached and resuspended in the culture medium and subsequently mixed with the xenograft in the ATSC group.

2.5. Cell Adhesion Capability

Cell adhesion was analyzed using an inverted optical microscope 3 days after seeding the cells into the culture flask, and it was verified that the cells were plastic adherent when maintained under the standard culture conditions.

2.6. Differentiation Assays

Adipogenic, osteogenic, and chondrogenic differentiation assays were done, following the same methodology adopted previously by our group (Coelho de Faria et al.) [7].

2.7. Immunophenotypic Characterization

The bone marrow-derived and adipose tissue-derived stem cells from the second passage were used for immunophenotypic characterization, following the same methodology used previously by our group (Coelho de Faria et al.) [7]. The cells showed compatible immunophenotyping (CD16+, CD34−, CD45−, CD73+, CD90+, and CD105+).

2.8. Seeding Cells into the Scaffolds

In all the groups, 1 mL of the solution containing phosphate buffered saline (PBS) and 1×10^5 cells were seeded into the scaffolds. The solution was slowly pipetted onto the scaffold which, due to its characteristic, was fully absorbed. At the end of the procedure, the scaffolds containing the cells inside were ready to be inserted in the surgical site.

2.9. Experimental Design and Surgical Protocol

Ten adult male New Zealand rabbits aged between 10 and 12 months and weighing between 3.5 and 4 kg were selected. The animals were acclimatized in individual cages for 14 days, in a temperature-controlled room (18 °C to 20 °C), subjected to a 12-h light cycle. The animals were fed a commercial pelleted diet and allowed ad libitum access to water.

All animals were subjected to anesthesia following the same general anesthesia protocol previously described. The animals received two commercial xenogenous blocks of bovine origin (Baumer, Mogi Mirim, SP, Brazil). The blocks were placed on their parietal bones, bilaterally to the midline with the aid of one fixing screw, after performing five perforations into the cortical plate of the parietal bone to encourage bleeding and graft nutrition (Figure 1).

The xenograft blocks from group BMSC were combined with the autologous bone marrow-derived mesenchymal stem cells (n = 10) and those from group ATSC were combined with the autologous adipose tissue-derived mesenchymal stem cells (n = 10); 1 mL of phosphate buffered saline (PBS) solution (Sigma Aldrich, Darmstadt, Germany) containing 1×10^5 cells was used in both groups, and the respective stem cells were added dropwise to each block. Each animal received one block of each group (i.e., BMSC group and ATSC group). The block from the BMSC group was fixed on the right side of the calvaria and the block from the ATSC Group was fixed on the left side of the calvaria. The surgical wounds were then subjected to primary closure.

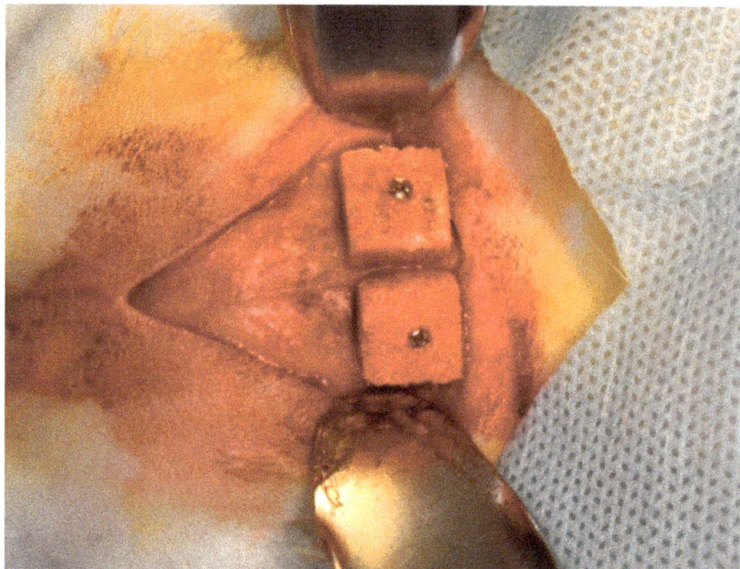

Figure 1. Xenograft blocks fixed on the rabbit's calvaria with titanium screws.

All animals were post-operatively medicated with benzylpenicillin (40,000 IU) and sodium dipyrone (0.25 mg/kg). The animals were sacrificed 8 weeks after surgery and their parietal bones were removed and processed for histological and histomorphometric evaluation.

2.10. Histological Preparation and Histomorphometric Evaluation

A portion of the parietal bone with an approximate area of 18 mm^2 containing the block at the center was removed using an oscillating saw. Samples were fixed in 10% formalin for 7 days and then decalcified in 10% EDTA for 6 days at room temperature and fixed with 10% buffered formaldehyde. The histological slices were prepared using a microtome to cut 5 μm transverse sections of the entire set, including the graft, from the center of the block (where the screw was positioned) up to 2 mm from the center of the block. The sections were stained with Masson's trichrome and assessed by optical microscopy (Nikon Eclipse C1, New York, NY, USA). A digital CCD camera was used to acquire images for the subsequent analyses (Infinity-1, Lumenera®, Ottawa, ON, Canada). All slides were analyzed in four areas (upper left, lower left, upper right, and lower right), which allowed the determination of tissue status in the interface, near the recipient bed (by using the lower left and lower right measurements) and also far from the recipient bed, at the block's surface (by using the upper left and upper right measurements). Then, the overall average was calculated for each slide. Each evaluated site had a dimension (area) of 1,347,442 μm^2. Two previously calibrated examiners assessed the specimens blindly and, in case of disagreement, the sample was reviewed to reach a consensus. An average of the measurements for each area obtained by the two examiners was recorded. The examiners traced all images using Infinity Analyse® (Lumenera Corporation, Ottawa, ON, Canada), measuring the following parameters: (1) newly formed bone (NFB), (2) xenogeneic residual particles (XRP), and (3) non-mineralized tissue (NMT). All results were obtained in μm^2 and expressed as percentage of the total area.

2.11. Statistical Analysis

Prior to the analyses, data from the newly formed bone (NFB), the xenogeneic residual particle (XRP) and the non-mineralized tissue (NMT) were evaluated for normality (Shapiro-Wilk tests) and homogeneity of variance (Levene tests), and it was verified that both were attended.

Analysis of variance with two criteria for randomized blocks (two-way ANOVA) was applied to investigate whether percentages of NFB, XRP, and NMT were influenced by independent variables of stem cell type (bone marrow/adipose tissue) and calvaria distance (near/far) or by the interaction of both. If there was a significant interaction, Tukey's test was used for multiple comparisons.

All quantitative data were analyzed by SPSS-V17® (SSPS Inc. 233, Chicago, IL, USA).

3. Results

Pearson's correlation test, which yielded a value of 0.99, was used to evaluate the measurements obtained from the two examiners. Therefore, we chose to use the mean measurements obtained by the two examiners. In low magnification (40×), the histological characteristics of the analyzed samples showed mineralized and non-mineralized tissues (Figure 2A). In higher magnification, both groups (BMSC and ATSC) showed areas of the newly formed bone, with a layer of osteoblasts adjacent to the reminiscent osseous trabeculae from xenograft blocks (Figure 2B,C, respectively). Variable quantities of the fibrovascular connective tissue were seen interspersed with the bone trabeculae (Figure 2B,C).

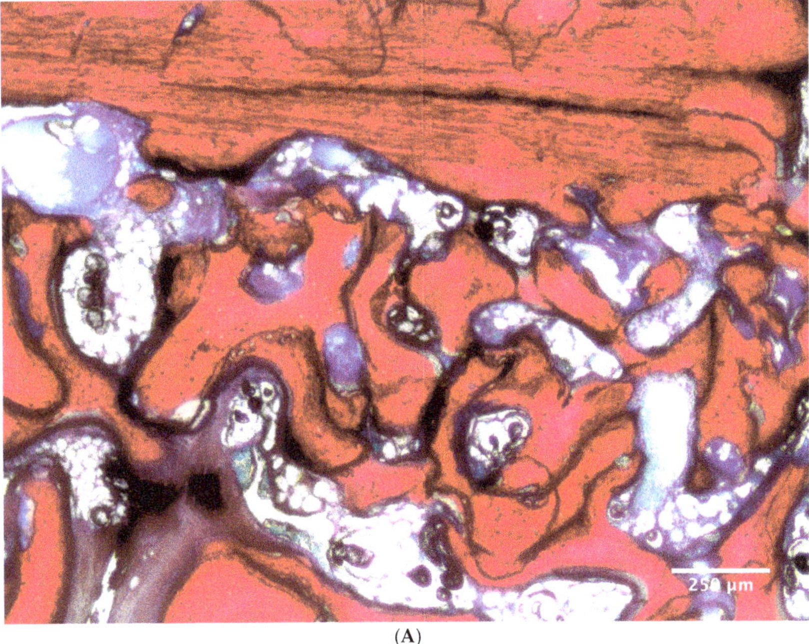

(A)

Figure 2. *Cont.*

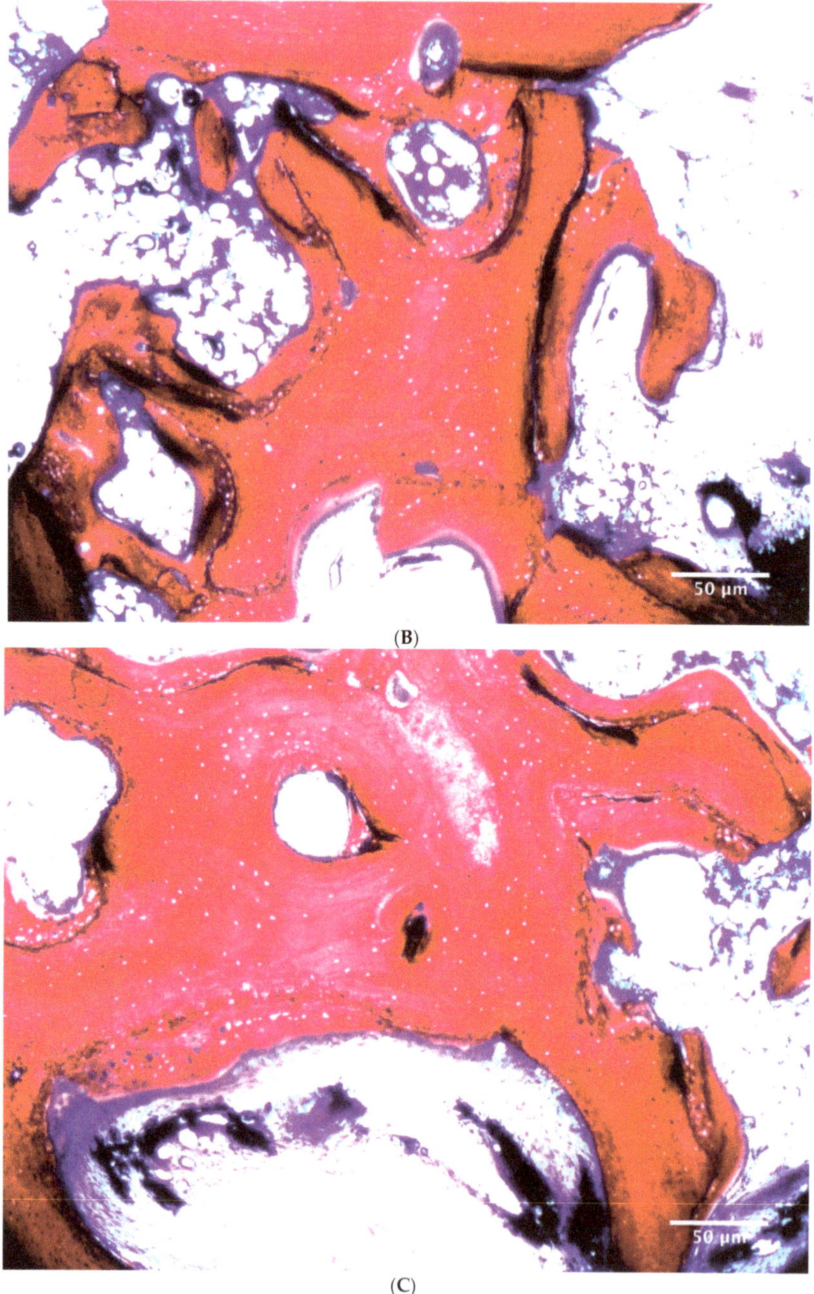

Figure 2. (**A**) Histologic characteristics of xenograft blocks combined with BMMSC (G1) and ATSC (adipose tissue stem cells) (G2). Histologic characteristics of a sample, in low magnification, showing native bone (above) and the grafted area, 40×. Stain, Masson's trichrome. (**B**) Higher magnification showing reminiscent bone trabeculae from G1 xenograft block surrounded by newly formed bone, 200×. Masson's trichrome. (**C**) Higher magnification showing reminiscent bone trabeculae from G2 xenograft block surrounded by the newly formed bone, 200×. Stain, Masson's trichrome.

For the newly formed bone (NFB) data, the two-way ANOVA showed no significant interaction between the concentrate type variables and the distance in the rabbit calvaria (p = 0.356). The results showed that the type of stem cell used significantly affected the percentage of the newly formed bone (NFB) (p = 0.036), a fact that occurred near (interface between the graft and the recipient bed) or far (graft's surface) from the recipient bed. Specifically, by Tukey's test, it was found that the impregnation of the xenograft block with the adipose tissue stem cells resulted in a percentage of newly formed bone (NFB) statistically lower than that observed for the group in which the bone marrow stem cells were used. Analysis of variance with two criteria for randomized blocks also indicated that in groups in which the xenograft block was impregnated with the adipose tissue stem cells or the bone marrow stem cells, the highest percentage of newly formed bone (NFB) was observed at a location near the recipient bed (p = 0.003).

For the percentage of xenogeneic residual particle (XRP), there was a significant interaction between both variables, type of cell and distance in rabbit calvaria (p = 0.063), as indicated by the analysis of variance with two criteria for randomized blocks. This test also identified no statistically significant difference (p = 0.750) between the groups (i.e., bone marrow and adipose tissue stem cells) in the percentage of xenogeneic residual particle (XRP). When the distances were compared, it was found that the percentage of xenogeneic residual particle (XRP) was significantly higher in the more distant location (i.e., far from the recipient bed) (p = 0.002).

Evaluating non-mineralized tissue (NMT), the two-way analysis of randomized blocks indicated a significant interaction between the stem cell type (i.e., BMSC and ATSC) and rabbit calvaria distance (i.e., near and far) (p = 0.038). Applying the Tukey's test, in close proximity to the calvaria, it was verified that the impregnation of the xenogeneic bone with the adipose tissue stem cells (ATSC) resulted in a greater percentage of non-mineralized tissue (NMT) in comparison to the group in which the impregnation occurred with the bone marrow stem cells (BMSC). In the more distant location, both groups did not differ significantly in relation to the percentage of non-mineralized tissue (NMT). The Tukey test also showed that the percentage of non-mineralized tissue (NMT) was higher in the distant location for the bone marrow stem cell group, whereas the adipose tissue stem cell group showed the highest percentage of non-mineralized tissue (NMT) at the nearest location. Table 1 shows the histomorphometric results and statistical comparisons.

Table 1. Histomorphometric results (mean and standard deviation) and statistical comparison between groups (in %)—BMSC, bone marrow stem cells group; ATSC, adipose tissue stem cells group; XRP, xenograft residual particles; NFB, newly formed bone; and NMT, non-mineralized tissue. Numbers inside the brackets are standard deviation and outside the brackets are mean. The mean values followed by different capital letters indicate statistically significant difference between the groups, within each column, considering separately each type of tissue. The mean values followed by different small letters indicate statistically significant difference between the sites, within each line, considering separately each type of tissue.

TISSUE	CELL SOURCE	DISTANCE-NEAR	DISTANCE-FAR	MEAN
NFB	BMSC	16.0 ± 6.1%	7.8 ± 6.7%	11.9 ± 7.5% A
	ATSC	9.9 ± 5.4%	5.3 ± 5.2%	7.6 ± 5.6% B
	MEAN	12.9 ± 6.4% a	6.6 ± 5.9% b	-
XRP	BMSC	20.7 ± 9.9%	24.1 ± 7.0%	22.14 ± 8.5% A
	ATSC	15.2 ± 4.9%	28.0 ± 6.3%	21.6 ± 8.5% A
	MEAN	17.9 ± 8.1% a	26.1 ± 6.8% b	-
NMT	BMSC	63.4 ± 10.5% Bb	68.3 ± 10.3% Aa	-
	ATSC	74.8 ± 7.4% Aa	66.7 ± 7.4% Ab	-

4. Discussion

There is consensus in the literature about the potential use of bone substitute materials to replace autogenous bone grafts in some clinical situations [20–22]. Nevertheless, in critical defects (e.g., vertical bone reconstruction), the lack of a cellular osteogenic component may limit the use of bone substitute biomaterials [11]. Therefore, the study of cell therapy is extremely important in order to remedy this situation.

As the adult mesenchymal stem cells have the capacity to differentiate into specialized cells of at least one mesenchymal lineage such as bone, cartilage, fat, or muscle [13], the use of these undifferentiated cells harvested from the adipose tissue and the bone marrow seems to have clinical applicability in regenerative medicine and, as far as the implant therapy is concerned, in bone reconstruction as well. The mesenchymal stem cells used in the present study fulfilled the minimal criteria proposed by the Mesenchymal and Tissue Stem Cell Committee of the International Society for Cellular Therapy (ISCT) position statement, which defined multipotent mesenchymal stromal cells [23]. These criteria were tested by cell adhesion, differentiation assays, and immunophenotypic analysis carried out as described in the Methods section. However, despite the fact the authors of the present study have used this standardization, SEM analysis was not performed to visualize the mesenchymal stem cells inside the scaffold.

In the present study, the percentage of newly formed bone was lower than in a previous study by our group [7], where we used a xenograft as a scaffold for the vertical bone construction on rabbit calvaria associated with the bone marrow-derived mesenchymal stem cells or the adipose tissue-derived stem cells. These differences in the levels of bone formation between the studies may have been caused by differences in (1) the characteristics of the scaffold and/or (2) the titanium device used to stabilize the scaffold. The presence of a particulate xenogenous graft in the first study could have contributed to a more adequate level of revascularization and, therefore, bone formation, when compared to the structured block graft of the present study. Moreover, the use of a titanium cylinder instead of a titanium screw certainly resulted in a larger titanium area in contact with the graft and bone in the previous study, which might have stimulated bone formation as titanium has a higher affinity for bone [24].

However, the XRP levels were similar between these two studies, with rates of 22.14 $\pm$ 8.5% and 21.6 $\pm$ 8.5% for the bone marrow-derived and the adipose tissue-derived mesenchymal stem cells, respectively, in the present study, and 23.31 $\pm$ 3.11% and 27.58 $\pm$ 3.98% for the bone marrow-derived and the adipose tissue-derived mesenchymal stem cells, respectively, in the previous study. Therefore, the use of a particulate or structured bone scaffold and the differences between the titanium areas probably did not have any influence. On the other hand, the NMT levels appear to be influenced by the scaffold characteristic and/or titanium area, since the percentage of this tissue in the group where the bone marrow-derived and the adipose tissue-derived stem cells were used was 65.8 $\pm$ 10.4% and 70.8 $\pm$ 7.4%, respectively, in the present study, and 50.23 $\pm$ 8.72% and 49.90 $\pm$ 8.76%, respectively, in the previous study. Therefore, the use of a particulate mineralized scaffold instead of a structured mineralized scaffold and the use of a larger area of titanium device to stabilize the scaffold may result in more newly formed bone and less soft tissue, which could contribute to an improved osseointegration when an implant is placed. However, it is important to state that the xenografts used in these two studies were not processed by the same company and, therefore, it might also contribute to different results.

Previous studies undertaken by our research group using the bone marrow-derived and the adipose tissue-derived mesenchymal stem cells in bone defects on rabbit calvaria showed the same tendency of higher levels of bone formation when using bone marrow stem cells (Pelegrine et al., 2014; Aloise et al., 2015; Zimmermann et al., 2015) [11,25,26]. Therefore, in both situations, onlay and inlay bone constructions, bone marrow may be considered a better choice for the tissue source of mesenchymal stem cells when compared to fat. The best results, in all studies, at the sites where the bone marrow-derived mesenchymal stem cells were used might be explained by the fact that cells from the bone marrow have a greater affinity for osteogenic differentiation, as stated before by our group [7]. Moreover, as mesenchymal stem cells are multipotent cells that are capable of multiple lineage differentiation due to the presence of inductive signals from the microenvironment [27,28], we hypothesize that the microenvironment of bone marrow present in the recipient bed had a more pronounced effect on the stem cells isolated from the bone marrow than on those obtained from the adipose tissue. This might also have repercussion in the higher levels of non-mineralized tissue at the interface between the recipient bed and the graft, as observed in the ATSC group. This finding represents a worse integration of the graft combined with the adipose tissue stem cells and, if proven by future clinical studies, could reflect in the implants' survival rates, as the dental implants are commonly installed in this interface area.

5. Conclusions

The bone marrow-derived mesenchymal stem cells seem to have a higher potential for bone formation compared with the adipose tissue-derived mesenchymal stem cells when used in combination with the xenogenous blocks in the vertical bone construction.

Author Contributions: Conceptualization, T.C.-A., A.A.P. and J.L.C.-G.; methodology, A.C.A.; software, T.C.-A.; validation, J.L.C.-G.; M.H.N.; formal analysis, A.C.A. and J.M.A.; investigation, C.E.S.d.C.; resources, A.C.A.; data curation, F.R.T.C.; writing—original draft preparation, F.P.-S.; writing—review and editing, F.R.T.C. and J.L.C.-G.; visualization, M.H.N.; supervision, A.A.P.; project administration, A.A.P.; and funding acquisition, A.A.P.

Funding: This study was financed in part by the Coordenação de Aperfeiçoamento de Pessoal de Nível Superior-Brasil (CAPES)-Finance Code 001.

Conflicts of Interest: The authors declare no conflict of interest.

References

1. Sakkas, A.; Wilde, F.; Heufelder, M.; Winter, K.; Schramm, A. Autogenous bone grafts in oral implantology-is it still a "gold standard"? A consecutive review of 279 patients with 456 clinical procedures. *Int. J. Dent.* **2017**, *3*, 23. [CrossRef] [PubMed]
2. Laurie, S.W.; Kaban, L.B.; Mulliken, J.B.; Murray, J.E. Donor-site morbidity after harvesting rib and iliac bone. *Plast. Reconstr. Surg.* **1984**, *73*, 933–938. [CrossRef] [PubMed]
3. Hernigou, P.; Desroches, A.; Queinnec, S.; Lachaniette, C.H.F.; Poignard, A.; Allain, J.; Chevallier, N.; Rouard, H. Morbidity of graft harvesting versus bone marrow aspiration in cell regenerative therapy. *Int. Orthop.* **2014**, *38*, 1855–1860. [CrossRef] [PubMed]
4. Monaco, E.; Bionaz, M.; Hollister, S.J.; Wheeler, M.B. Strategies for regeneration of the bone using porcine adult adipose-derived mesenchymal stem cells. *Theriogenology* **2011**, *75*, 1381–1399. [CrossRef]
5. Fu, X.; Jin, L.; Ma, P.; Fan, Z.; Wang, G.S. Allogeneic stem cells from deciduous teeth in treatment for periodontitis in miniature swine. *J. Periodontol.* **2014**, *85*, 845–851. [CrossRef] [PubMed]
6. Yamada, Y.; Nakamura, S.; Ueda, M.; Ito, K. Papilla regeneration by injectable stem cell therapy with regenerative medicine: Long-term clinical prognosis. *J. Tissue Eng. Regener. Med.* **2015**, *9*, 305–309. [CrossRef] [PubMed]
7. Coelho de Faria, A.B.C.; Chiantia, F.B.; Teixeira, M.L.; Aloise, A.C.; Pelegrine, A.A. Comparative study between mesenchymal stem cells derived from bone marrow and from adipose tissue, associated with xenograft, in appositional reconstructions: Histomorphometric study in rabbits calvaria. *Int. J. Oral Maxillofac. Implants* **2016**, *31*, e155–e161. [CrossRef] [PubMed]
8. Katagiri, W.; Osugi, M.; Kawai, T.; Hibi, H. First-in-human study and clinical case reports of the alveolar bone regeneration with the secretive from human mesenchymal stem cells. *Head Face Med.* **2016**, *15*, 12–15.
9. Zhu, X.; Liu, J.; Yu, Z.; Chen, C.A.; Aksel, H.; Azim, A.A.; Huang, G.T. A Miniature Swine Model for Stem Cell-Based De Novo Regeneration of Dental Pulp and Dentin-Like Tissue. *Tissue Eng. Part. C Methods* **2018**, *24*, 108–120. [CrossRef]
10. Meschi, N.; Hilkens, P.; Lambrichts, I.; Van den Eynde, K.; Mavridou, A.; Strijbos, O.; De Ketelaere, M.; Van Gorp, G.; Lambrechts, P. Regenerative endodontic procedure of an infected immature permanent human tooth: An immunohistological study. *Clin. Oral. Investig.* **2016**, *20*, 807–814. [CrossRef]

11. Pelegrine, A.A.; Aloise, A.C.; Zimmermann, A.; de Mello e Oliveira, R.; Ferreira, L.M. Repair of critical-size bone defects using bone marrow stromal cells: A histomorphometric study in rabbit calvaria. Part I: Use of fresh bone marrow or bone marrow mononuclear fraction. *Clin. Oral Implants Res.* **2014**, *25*, 567–572. [CrossRef] [PubMed]
12. Victorelli, G.; Aloise, A.C.; Passador-Santos, F.; de Mello e Oliveira, R.; Pelegrine, A.A. Ectopic Implantation of Hydroxyapatite Xenograft Scaffold Loaded with Bone Marrow Aspirate Concentrate or Osteodifferentiated Bone Marrow Mesenchymal Stem Cells: A Pilot Study in Mice. *Int. J. Oral. Maxillofac. Implants* **2016**, *31*, e18–e23. [CrossRef] [PubMed]
13. Chamberlain, G.; Fox, J.; Ashton, B.; Middleton, J. Concise review: Mesenchymal stem cells: Their phenotype, differentiation capacity, immunological features, and potential for homing. *Stem Cells* **2007**, *25*, 2739–2749. [CrossRef] [PubMed]
14. Kawaguchi, H.; Hirachi, A.; Hasegawa, N.; Iwata, T.; Hamaguchi, H.; Shiba, H.; Takata, T.; Kato, Y.; Kurihara, H. Enhancement of periodontal tissue regeneration by transplantation of bone marrow mesenchymal stem cells. *J. Periodontol.* **2004**, *75*, 1281–1287. [CrossRef] [PubMed]
15. Moseley, T.A.; Zhu, M.; Hedrick, M.H. Adipose-derived stem and progenitor cells as fillers in plastic and reconstructive surgery. *Plast. Reconstr. Surg.* **2006**, *118* (Suppl. 3), 121S–128S. [CrossRef] [PubMed]
16. Machado, E.; Fernandes, M.H.; de Gomes, P.S. Dental stemcells for craniofacial tissue engineering. *Oral Surg. Oral Med. Oral Pathol. Oral Radiol.* **2012**, *113*, 728–733. [CrossRef] [PubMed]
17. Mitrano, T.I.; Grob, M.S.; Carrion, F.; Nova-Lamperti, E.; Luz, P.A.; Fierro, F.S.; Quintero, A.; Chaparro, A.; Sanz, A. Culture and characterization of mesenchymal stem cells from human gingival tissue. *J. Periodontol.* **2010**, *81*, 917–925. [CrossRef]
18. Man, Y.; Wang, P.; Guo, Y.; Xiang, L.; Yang, Y.; Qu, Y.; Gong, P.; Deng, L. Angiogenic and osteogenic potential of platelet-rich plasma and adipose-derived stem cell laden alginate microspheres. *Biomaterials* **2012**, *33*, 8802–8811. [CrossRef]
19. Kaully, T.; Kaufman-Francis, K.; Lesman, A.; Levenberg, S. Vascularization of the conduit to viable engineered tissues. *Tissue Eng. Part B Rev.* **2009**, *15*, 159. [CrossRef]
20. Araujo, D.B.; de Jesus Campos, E.; Oliveira, M.A.; Lima, M.J.; Martins, G.B.; Araujo, R.P. Surgical elevation of bilateral maxillary sinus floor with a combination of autogenous bone and lyophilized bovine bone. *J. Contemp. Dent. Pract.* **2013**, *14*, 445–450. [CrossRef]
21. Dottore, A.M.; Kawakami, P.Y.; Bechara, K.; Rodrigues, J.A.; Cassoni, A.; Figueiredo, L.C.; Piattelli, A.; Shibli, J.A. Stability of implants placed in augmented posterior mandible after alveolar osteotomy using resorbable nonceramic hydroxyapatite or intraoral autogenous bone: 12-month follow-up. *Clin. Implant Dent. Relat. Res.* **2014**, *16*, 330–336. [CrossRef]
22. Pang, C.; Ding, Y.; Zhou, H.; Qin, R.; Hou, R.; Zhang, G.; Hu, K. Alveolar ridge preservation with deproteinized bovine bone graft and collagen membrane and delayed implants. *J. Craniofac Surg.* **2014**, *25*, 1698–1702. [CrossRef]
23. Dominici, M.; Le Blanc, K.; Mueller, I.; Slaper-Cortenbach, I.; Marini, F.; Krause, D.; Deans, R.; Keating, A.; Prockop, D.J.; Horwitz, E. Minimal criteria for defining multipotent mesenchymal stromal cells. The International Society for Cellular Therapy position statement. *Cytotherapy* **2006**, *8*, 315–317. [CrossRef]
24. Brånemark, P.I. Osseointegration and its experimental background. *J. Prosthet. Dent.* **1983**, *50*, 399–410. [CrossRef]
25. Aloise, A.C.; Pelegrine, A.A.; Zimmermann, A.; de Mello e Oliveira Ferreira, L.M. Repair of Critical-Size Bone Defects Using Bone Marrow Stem Cells or Autogenous Bone with or Without Study in Rabbit Calvaria. *Int. J. Oral Maxillofac. Implants* **2015**, *30*, 208–215. [CrossRef] [PubMed]
26. Zimmermann, A.; Pelegrine, A.A.; Peruzzo, D.; Martinez, E.F.; de Mello e Oliveira, R.; Aloise, A.C.; Ferreira, L.M. Adipose mesenchymal stem cells associated with xenograft in a guided bone regeneration model: A histomorphometric study in rabbit calvaria. *Int. J. Oral Maxillofac. Implants* **2015**, *30*, 1415–1422. [CrossRef] [PubMed]

27. Mao, J.J.; Giannobile, W.V.; Helms, J.A.; Hollister, S.J.; Krebsbach, P.H.; Longaker, M.T.; Shi, S. Craniofacial tissue engineering by stem cells. *J. Dent. Res.* **2006**, *85*, 966–979. [CrossRef]
28. Zhang, L.; Feng, G.; Wei, X.; Huang, L.; Ren, A.; Dong, N.; Wang, H.; Huang, Q.; Zhang, Y.; Deng, F. The effects of mesenchymal stem cells in craniofacial tissue engineering. *Curr. Stem Cell Res. Ther.* **2014**, *9*, 280–289. [CrossRef]

Article

Evaluation of the Cortical Deformation Induced by Distal Cantilevers Supported by Extra-Short Implants: A Finite Elements Analysis Study

Enrique Fernández-Bodereau [1,*], Viviana Yolanda Flores [2], Rafael Arcesio Delgado-Ruiz [3], Juan Manuel Aragoneses [4] and José Luis Calvo-Guirado [5]

1 Department of Fixed Prosthodontics, Faculty of Dentistry, Universidad Nacional de Córdoba, Córdoba 5000, Argentine

2 Department of Oral Biology, Faculty of Dentistry, Universidad Nacional de Córdoba, Córdoba 5000, Argentine; vivyflores@gmail.com

3 Department of Prosthodontics, School of Dental Dental Medicine, Stony Brook, NY 1103, USA; Rafael.Delgado-Ruiz@stonybrookmedicine.edu

4 Department of Dental Research in Universidad Federico Henríquez y Carvajal (UFHEC), Santo Domingo 10107, Dominicana Republic; jmaragoneses@gmail.com

5 Department of Oral and Implant Surgery, Faculty of Health Sciences, Universidad Católica San Antonio de Murcia (UCAM), Murcia 30002, Spain; jlcalvo@ucam.edu

* Correspondence: fernandezenrique5@gmail.com

Received: 10 November 2018; Accepted: 11 December 2018; Published: 17 December 2018

Abstract: Background: The aim of the study was to analyze the distribution of stresses caused by an axial force in a three-dimensional model with the finite element method in the implant-supported fixed partial denture with distal overhang (PPFIVD) on short dental implants in the posterior edentulous maxilla. Methods: geometrical models of the maxilla with a bone remnant of 9 and 5 mm were created. Straumann SP® (Base, Switzerland) implants were placed in the premolar area. Two groups with subgroups were designed. Group A (GA): PPFIVD on two implants (GA1: 4.1 × 8 mm and GA2: 4.1 × 4 mm); Group B (GB): PPFIVD on the single implant (GB1: 4.1 × 8 mm and GB2: 4.1 × 4 mm). It was applied to a static force of 100 N to 30°. Results: PPFIVD on two implants reached the maximum tension in GA2 with respect to GA1; the difference was not significant in implants. In the maxilla GA2 was lower in relation to GA1; the difference was not significant. In PPFIVD over an implant, the stress was greater in GB2 with respect to GB1; the difference was significant in maxilla and implants. Peri-implant bone micro deformations and prosthesis-implant displacements were observed. Conclusions: PPFIVD over short splinted implants could be viable in the maxilla with reduced bone height, being an option when lifting the floor of the maxillary sinus. The rehabilitation with unitary implant (4 mm) did not provide adequate results. The dominant tensions evidenced bone micro-distortions with a displacement of the prosthesis-implant set. The real statement of this paper was to define that short splinted implants can be used in soft bone with high success rate in reducing bending forces.

Keywords: short dental implant; finite element analysis; distal cantilever; implant-supported partial fixed prosthesis; microdeformations; displacements; geometric model; maxilla

1. Introduction

The Brånemark system defined osseointegration as the direct bonding between titanium implants and the surrounding bone, maintained during functional loads [1,2]. After extraction of the teeth in the posterior edentulous maxilla, the resorption of the bony crest, the pneumatization of the maxillary sinus, or the combination of both are evidenced, which reduces the possibility of placing fixings of standard length and is generally accompanied by complex surgical procedures [3,4]. In general, these surgical treatments include: maxillary floor elevation and grafts, other procedures combine grafting and the use of membranes [5,6]. An alternative technique to complex surgical procedures is the use of short implants (implants with lengths of $\leq$8 mm) [7,8].

Recently, extra-short implants of 4 mm in length have emerged. Extra-short implants are destined to zones with noticeable bony resorption unable to support a short implant, thus, avoiding the use invasive techniques, potential reduction of the time of healing and the financial burden for the patients. In the case of using extra-short implants, the crown/implant ratio can be of 2–3:1 varying with the prosthetic space [9,10].

During the mechanical load of the implant restoration, adaptive processes occur at the peri-implant supporting tissues that influence the healing, remodeling, and resorption of the surrounding bone [11,12]. Physiological bone deformation happens between 1000 $\mu\varepsilon$ and 1500 $\mu\varepsilon$ ($\mu\varepsilon$ = microstrains) in the normal bite [13–15]. However, in extra-short implants, these values may increase, affecting bone healing and regeneration. [14,16].

Indeed, measurements of static bite force indicate that magnitude is in average 100–150 N in adult men [17,18]; the location of the implants in the edentulous maxilla determines the force intensity, being the posterior areas those with the higher registered forces [19]. Of preference, a balanced occlusion can grant a better distribution of forces [20].

Important parameters were established for the use of cantilevers with standard and long implants [21]. However, there is a lack of information of the long-term effects cantilevers supported by extra-short implants [22–24]. Finite element analysis (FEA) has been used to evaluate the biomechanics of extra-short implants under axial forces [2,14,15].

The hypothesis of the present study is that that the forces generated during the mastication in the posterior region of the maxilla transmitted through the prosthetic restoration with a distal cantilever to the standard and extra-short implants could affect the osseous tissue in a similar fashion.

The objective of the present research was to analyze with FEA, the distribution of tensions produced by an axial force applied to the fixed partial prosthesis with distal cantilever, supported by extra-short implants of 4 mm length in the posterior sector of the partially edentulous maxilla.

2. Materials and Methods

For this experimental FEA study, the following variables considered: Standard Plus® (Straumann, Basel, Switzerland) implants with a diameter of 4.1 mm by 8 mm in length and extra-short implant with a diameter of 4.1 mm by 4 mm in length.

The behavior of the bone and the prosthesis under simulated static load was evaluated. The distribution of the tensions was studied as per ISO 14801:2003 and ISO 1942-1 standards. The geometric model of a unilateral edentulous maxilla was generated using Abaqus®, 6.4 Version (Abaqus Inc., Providence, RI, USA) through the use of optical scanners [25].

The distribution of the tensions in bone when applying an axial force on the fixed partial prosthesis was also evaluated. The geometry of a partially edentulous maxillary segment was determined with a frontal cut around the first premolar; dimensions in the three axes of the space were registered.

The dimensions of the implant, abutment, and prosthetic components were settled down with a digital caliper; a representative average with the projection of the profiles was registered, using a TAF003 profilometer from Mitutoyo® (Tokyo, Japan).

The coefficient and modulus of elasticity (e) of Poisson (V) were assigned as follows: Cortical bone (E13700 MPa; V0.3), trabecular bone (E13700 MPa; V0.3), titanium implants (E110000 MPa; V0.3), and chromium-nickel prosthesis (E171000 MPa; V0.3) did not include the cement because its modulus of elasticity is low with respect to the used materials [13,26]. The geometric model was characterized as a trabecular core surrounded by a cortical layer that resembles the characteristics of a partially toothless maxilla [27].

Available bone heights were 9 and 5 mm. In the edentulous section of the first and second premolar and first molar, short and extra-short implants were inserted, a connection of the abutments to the fixed partial denture with the distal cantilever (PPFIVD) was simulated. The mesiodistal distance of each cantilever averages 7 mm.

Two groups were analyzed: group A (GA) constituted by two subgroups: GA1, two short implants of 4.1 × 8 mm; and GA2, two extra-short implants of 4.1 × 4 mm. Group B (GB), also had two subgroups: GB1, a short implant of 4.1 × 8 mm, and GB2, with an extra short implant of 4.1 × 4 mm.

The osseous tissue was modeled in close relation with the implant threads to accentuate the precision of the results (Table 1) with a mechanically ideal interface and assuming 100% of bone-to-implant contact.

Table 1. Numeric detail geometric patterns made for this investigation.

Models	Amount of Elements	Number of Nodes	Types of Elements	Bone Height (mm)	Implants Dimensions (mm)
Two implants	321,066	65,826	Tetrahedron	9	4.1 × 8
Two implants	424,672	85,442	Tetrahedron	5	4.1 × 4
An implant	199,443	40,776	Tetrahedron	9	4.1 × 8
An implant	264,076	53,261	Tetrahedron	5	4.1 × 4

Algorithmic calculations were applied to define the geometry, to generate the mesh, the load conditions, and the properties of the materials. Afterward, a set of n-equations and n-incognito were solved with an algorithm for the system [26]. Linear homogeneous and isotropic equations were applied. With all the registered data, magnitudes derived from the nodes were calculated (Table 1).

A physiological static load of 100 N (Newton) with an angulation of 30° was applied [18,19] by means of a standard circular piece whose purpose was to distribute the load uniformly. The application point was standardized for all the models in mesial of the triangular crest of the palatal peak.

The present investigation was completed on the base of the criterion of three-dimensional tensions of Von Mises, expressed in the following formula:

σ_ (VM √ (((σ_1 〖σ〗_2) ^2 (σ_ (2) - σ_3) ^2 (σ_3 σ_1) ^2)/2))

where σVM is the tension of Von Mises and σ1; σ2; σ3 the principal stresses.

3. Results

3.1. Distribution of the Tensions in the Implant

In GA1 and GA2, the load applied in the crown it distributed the principal stresses in the cervical third of both implants; in the cantilever area, the tension increased at the cervical third of both implants, also including the middle third the second implant located at the level of the second premolar.

GA2 was greater with respect to GA1, but without significant difference. The distribution of the tensions in GB1 and GB2 was observed in the cervical third of both implants in the cantilever, the principal stresses included the cervical third and middle in both sub-groups. GB2 was greater in relation to GB1 (Table 2; Figures 1–3).

Table 2. Distribution of the principal stresses in ultrashort implants.

Models	Crown (MPa)	Cantilever (MPa)	Load (N)	Implant Dimensions (mm)
A Two implants	183.7	173.9	100	4.1 × 8
	170	205.4	100	4.1 × 4
B An implant	330.2	334	100	4.1 × 8
	321.6	381.4	100	4.1 × 4

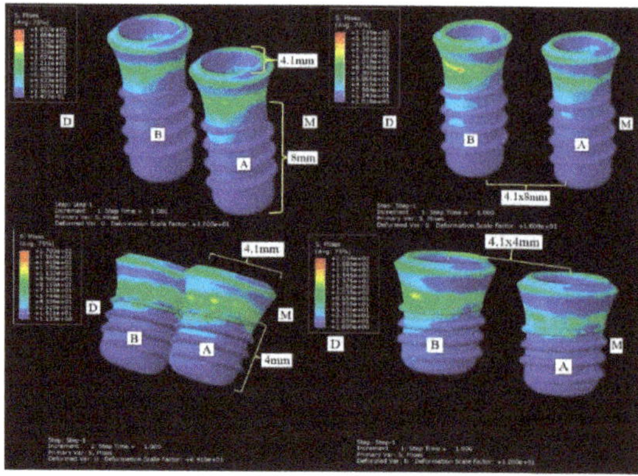

Figure 1. Distribution of the main stresses in short implants. On the left of the figure, the load on the crown of the implant-supported fixed partial prosthesis on two implants of 4.1 × 8 mm (top) and 4.1 × 4 mm (below) observed. To the right of the figure, the load in the cantilever is observed. A. Implant located at the level of the first upper premolar and B at the level of the second upper premolar. Cervical tensions are observed in both A and B.

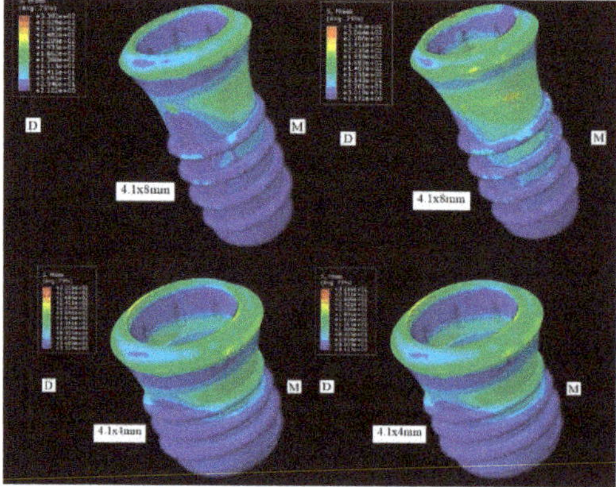

Figure 2. Distribution of the main tensions in the short implants located at the level of the first upper premolar. On the left of the figure, the load on the crown of the implant-supported fixed partial prosthesis is observed on an implant of 4.1 × 8 mm (on top) and 4.1 × 4 mm (below). To the right of the figure, the load in the cantilever is observed. Tensions are observed in the cervical and middle third.

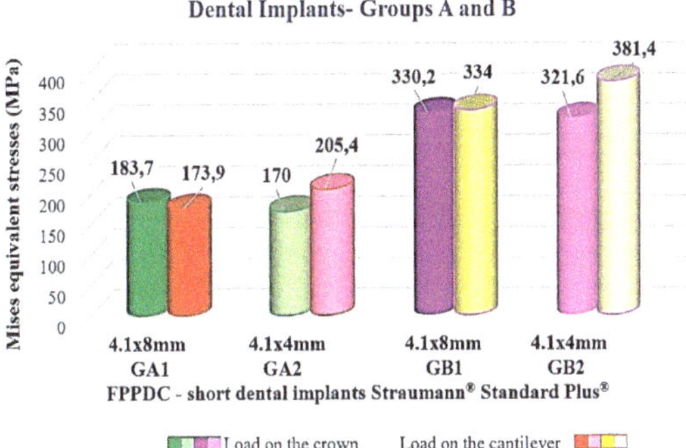

Figure 3. Distribution principal stresses in implants. Fixed partial prosthesis with distal cantilever, supported by ultrashort implants -FPPDC-.

3.2. Distribution of the Tensions in the Solid Model

The load in the crown distributed the tensions in the cervical third of GA1 and GA2; in the cantilever, tensions were observed in the cervical third corresponding to the first premolar, and in the cervical and middle third of the second premolar, in both sub-groups. GA1 was greater with respect to GA2, without significant difference among them.

In GB1 and GB2 the distribution of the principal stresses was demonstrated in the cervical third; in the cantilever, increased tensional values were observed in the cervical and middle third, in both sub-groups. GB2 was larger than GB1, with significant differences among them, marking a critical situation on a map for GB2 (Table 3; Figure 4).

Table 3. Distribution of the principal stresses in the maxillary.

Models	Crown (MPa)	Cantilever (MPa)	Load (N)	Implant Dimensions (mm)
A Two implants	57.55	86.26	100	4.1 × 8
	58.61	78.42	100	4.1 × 4
B An implant	96.83	128.5	100	4.1 × 8
	101.3	168.5	100	4.1 × 4

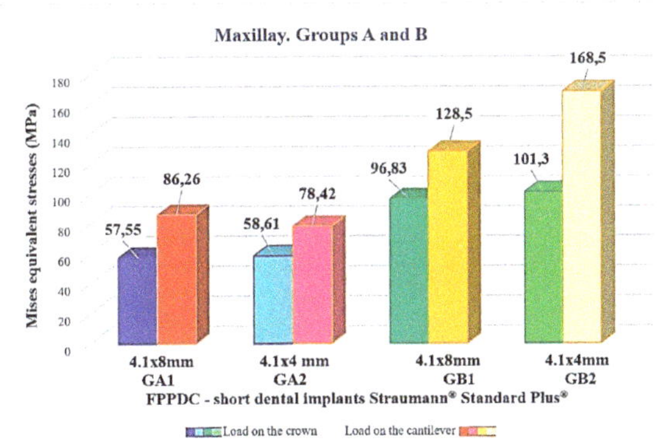

Figure 4. Distribution principal stresses in maxillary. Fixed implant-supported partial prosthesis with distal cantilever -FPPDC-.

3.3. Bone Behavior

The high tensions brought bone micro deformations in a proportional way (higher tensions then more deformation). In GB2 there were higher values with respect to GB1, with significant differences (Table 4 and Figures 5–9).

Table 4. Distribution of the microstrain in the solid model. Fixed partial prosthesis with distal cantilever, supported by ultrashort implants.

Models	Crown (Maximum Tensile) με	Cantilever (Maximum Tensile) με	Load (N)	Implants Dimensions (mm)
A Two implants	3.439	6.983	100	4.1 × 8
	6.648	5.990	100	4.1 × 4
B An implant	5.777	8.553	100	4.1 × 8
	6.538	9.253	100	4.1 × 4

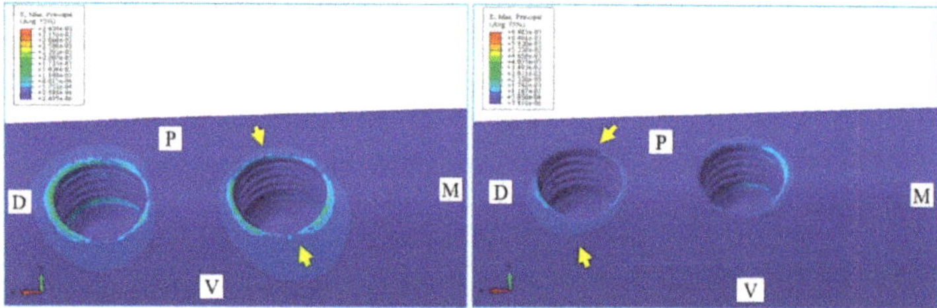

Figure 5. Microdeformations in the maxillary. Load on the crown (**left**) and the cantilever (**right**) of PPDC supported by two ultrashort implants of 4.1 × 8 mm. Arrows indicate microstrain in the bone crest.

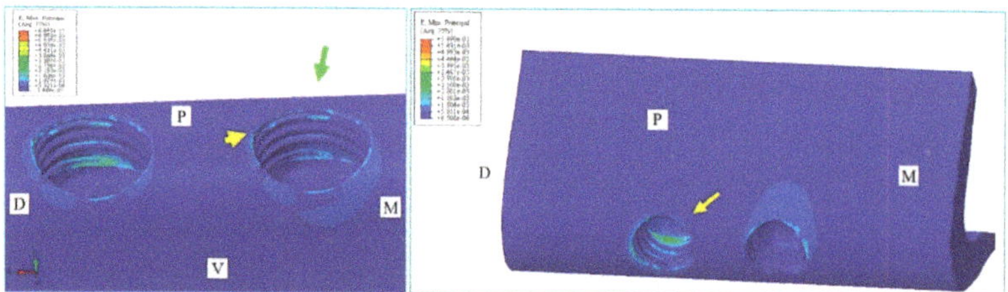

Figure 6. Microstrain in the maxillary. Load on the crown (**left**) and the cantilever (**right**) of the PPFIVD supported by two ultrashort implants of 4.1 × 4 mm. Arrows indicate microstrain in the bone crest.

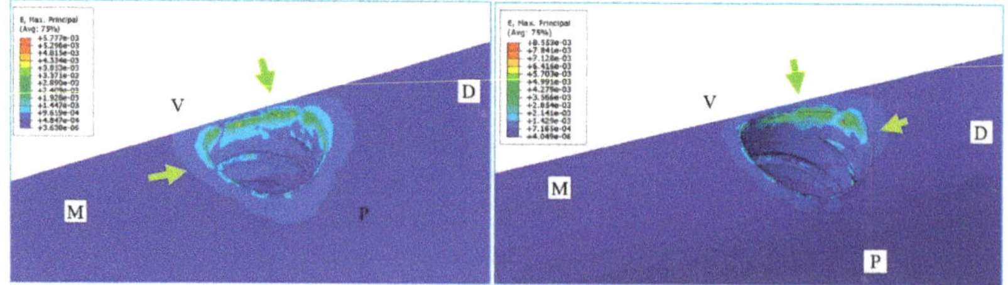

Figure 7. Microdeformations in the maxilla. Load on the crown (**left**) and on the cantilever (**right**) of the PPFIVD supported by an ultrashort implant of 4.1 × 8 mm. Arrows indicate microstrain in the bone crest.

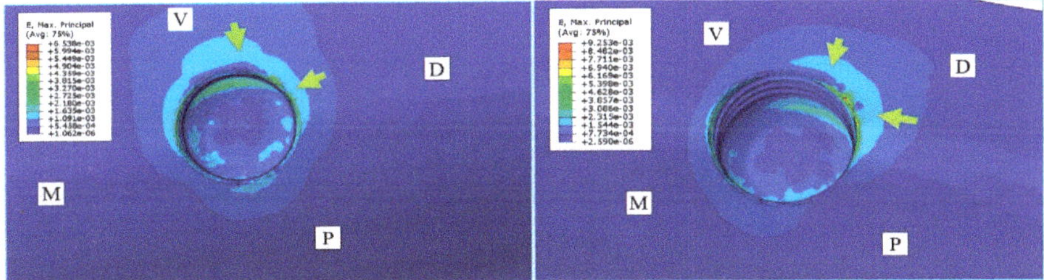

Figure 8. Microdeformations in the maxilla. Load on the crown (**left**) and the cantilever (**right**) of the PPFIVD supported by an ultrashort 4.1 × 4mm. Arrows indicate microstrain in the bone crest.

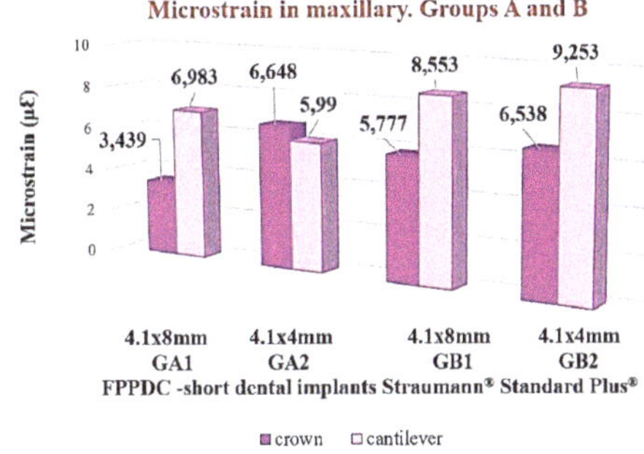

Figure 9. Microstrain in maxillary. Fixed partial prosthesis with distal cantilever, supported by ultrashort implants -FPPDC-.

3.4. Prosthesis Behavior

The micro-deformations facilitated the displacement of the joint prosthesis-implant. In GA1 and GA2 it was oriented through the buccal side. In GB1 and GB2 the displacement was in the buccal-distal direction. GB2 was a major that GB1, with significant difference (Table 5; Figure 10).

Table 5. Displacements of the fixed partial prosthesis with distal cantilever supported by ultrashort implants.

Models	Displacement Crown (mm)	Displacement Cantilever (mm)	Load (N)	Implant Dimensions (mm)
A Two short implants	0.02897	0.04072	100	4.1 × 8
	0.02896	0.04045	100	4.1 × 4
B A short implant	0.05933	0.00728	100	4.1 × 8
	0.06705	0.08347	100	4.1 × 4

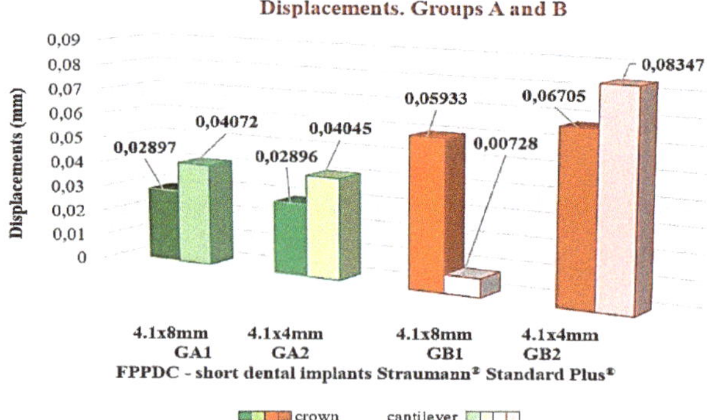

Figure 10. Displacements of the fixed partial prosthesis with distal cantilever, supported by ultrashort implants -FPPDC-.

4. Discussion

The obtained data were described in agreement with a Ceá hypothesis, the error committed in the approach of an exact solution was limited by the error of approach of the obtained solution [27]. The approach error depended crucially on the size and amount of the elements and the equality of other factors that caused this to be smaller, and it was limited to the error of the solution of finite elements. The distribution of short implants in the posterior sector of the maxilla is a challenge when the available bony height is reduced and is next to the floor of the maxillary sinus.

The tension of Von Mises is a physical magnitude of the distortion energy, and is a compound tension that indicates to the average level of tension whose maximum value indicates the possibility of damage occurring. The variables can be manipulated; the software eliminates the variation of the resulting probability of the error of the sample. The initial deformation of the majority of solids is elastic, which implies that the deformation is reversible when stopping applying the tension, and the reason why the solid recovers its initial form.

In the majority of cases, and in agreement with the Hooke's Law, the relation tension-deformation in the elastic regime is linear, and it expresses $\sigma = E \cdot \varepsilon$, where σ is the tension expressed in MPa and is the Young's modulus or modulus of elasticity, and ε is the deformation. If the modulus of elasticity is sprightly, the following formula is obtained whose obtained value indicates that the piece in the study can reconstitute its original form: $E = \sigma/\varepsilon$. However, when the maximum tensional values surpass the reversible capacity of the material it produces a permanent dimensional change. The deformation formula can be established simply: $\varepsilon = \sigma/E$. However, the three-dimensional resolution is more complex.

Sotto Maior et al. [28], described the stress concentration in short single implants of 5 mm in diameter by 7 mm in length in 32 partially-edentulous atrophic jaws. They simulated traumatic occlusion and a crown-implant relation of 2:1 or 2.5:1. The crown created high bending stresses which produced a peri-implant bone loss. Bolle et al. [29] used implants 4 mm in diameter and 10 mm in length over the mandibular channel with and without bone augmentation. Additionally, the inserted implants were 4 to 5 mm in length below the maxillary sinus with and without sinus elevation. After one year of load, the short implants obtained similar results compared to implants of conventional lengths (10 mm). They concluded that short implants could be an option to the surgical procedures since the treatment is less invasive, faster, and economical.

Toti et al. [30] inserted short implants in the posterior atrophic zone of partially-edentulous jaws, and their results showed better compared with implants of standard length placed in the augmented bone. Additionally, the short implants presented less crestal bone loss. Tabrizi et al. [31] evaluated the bone loss in 65 short implants in the jaw. There were no significant differences in the marginal bone loss between the three groups analyzed. However, the authors found that increasing the number of short implants reduced the percentages of crestal bone loss given a better force distribution. Gastaldi et al. [32] compared prosthesis on short implants of 5 × 5 mm or of 5 × 10 mm. inserted in atrophic jaws with augmented bone and maxillary bilateral sinus elevation. After three years of load, short implants obtained results similar to standard implants. They concluded that short implants can be a preferable option to the vertical bone augmentation.

Chou et al. [33], compared the stress distribution around the peri-implant bone of narrow implants with standard length (3.5 mm wide by 10.7 mm length) compared with short and wide implants (5 mm diameter by 7 mm length) in the mandible. It was demonstrated that the development of tension in the cervical region is inevitable, with more uniform distributions in the peri-implant tissue in the short and wide implants. Sahrmann et al. [34], compared implants 6 and 10 mm in length through X-rays. Higher mineralization percentages were observed around short implants compared to standard implants. This was related to the capacity of bone adaptation.

Peixoto et al. [35] used digital images to calculate the deformations of the prosthesis supported by splinted short dental implants during an occlusal load of 250 N. It was found that the veneering material did not influence the distribution of the deformation. Esfahrood et al. [36] performed a prospective study comparing the fixed prosthesis with overdentures supported by short implants (<10 mm). The cumulative success in the implant with rough surfaces was higher than in implants with machined surfaces. The survival of short implants in edentulous maxilla was high when the implants were placed under strict clinical protocols. Therefore, the authors consider the short implants as a safe and predictable technique.

On the other hand, Lemos et al. [37] compared short implants (≤8 mm) with standard implants (>8 mm) placed in the posterior regions of the maxilla and the mandible and found that shorter implants >8 mm presented a greater risk of failure. Block et al. [38] analyzed the level of bone deformation when changing the prosthetic design and different occlusal schemes. It was found that the restorations with splinted implants reduced the levels of deformation in the area of the neck of the implants and provide a better deformation distribution under cyclic load. De Souza Batista et al. [39] compared the influence of the pontic designs and the effect of mesial and distal cantilevers in prostheses supported by standard implants (4 × 10 mm). Axial forces of 400 N and of 200 N were applied in an oblique direction. In the group with two implants supporting a distal cantilever, an unfavorable biomechanical behavior with higher stresses was observed.

In the present work, there were no observed differences between GA1 and GA2 implants. The tensions in GB2 were, principally, greater than those of GB1. However, GB2 (4.1 × 4 mm) exerted higher tensions, an unpredictable situation for implants of short height. Ormianer et al. [40] studied the marginal bone loss related to implants of different designs, (different threads, thread direction, and apical design). In the long-term, there was minor bone loss in implants with greater interthread distances, deep apical turns, and narrow cores. Their results favored a single-piece implant design with a "V" thread design. Wolff et al. [14] demonstrated that the effective tensions take place in the bone–implant interface under unidirectional axial load. Four types of implants showed bone loss resulting from pathological tensions >3000 με. The groups analyzed in the present work were in the rank of pathological overload; this favored the micro deformations and the displacements of the implant/abutment connections.

Villarinho et al. [41] inserted 46 single implants of 4.1 × 6 mm in the posterior areas of the maxilla and the mandible in 20 patients; after 45 ± 9 months the percentages of success were 65.2% and 95%, with more probabilities of implant loss in the jaw than in the maxilla. The crown–implant ratio and the time were significant predictive values for bone loss ($p < 0.001$). In this research that GA two splinted short implants were observed, which would have the best stress distribution and, therefore, more possibility of permanence in the mouth, and which would give the joint prosthesis-implant greater stability in the long-term.

5. Conclusions

Within the limitations of this FEA:

- PPFIVD on two extra-short dental implants could be used as a viable option;
- The tensions are distributed in the cervical third of the implants and to the implant/abutment interface in the form of micro deformations;
- PPFIVD over single extra short implants of 4.1 × 4 mm possess a high risk or mechanical failure; and
- The forces generated during the functional load in the posterior area of the maxilla would affect the prosthetic restoration with distal cantilever and would affect the peri-implant bone that surrounds the extra-short dental implants.

Author Contributions: Conceptualization, Enrique Fernández Bodereau.; methodology, Rafael Arcesio Delgado Ruiz, Viviana Yolanda Flores.; software, Viviana Yolanda Flores.; validation, Enrique Fernández Bodereau. and Viviana Yolanda Flores.; formal analysis, Juan Manuel Aragoneses, Viviana Yolanda Flores.; investigation, Viviana Yolanda Flores.; resources, José Luis Calvo Guirado.; data curation, Viviana Yolanda Flores.; writing—original draft preparation, Rafael Arcesio Delgado Ruiz, José Luis Calvo Guirado.; writing—review and editing, Enrique Fernández Bodereau and Viviana Yolanda Flores.; visualization, Juan Manuel AragonesesViviana Yolanda Flores and José Luis Calvo Guirado.; supervision, Enrique Fernández Bodereau.; project administration, Enrique Fernández Bodereau.; funding acquisition, Enrique Fernández Bodereau."

Funding: This research received no external funding.

Acknowledgments: To the engineer Domínguez A., for his technical collaboration in obtaining the consigned data.

Conflicts of Interest: The authors declare no conflict of interest

References

1. Brånemark, P.I. Osseointegration and its experimental background. *J. Prosthet. Dent.* **1983**, *50*, 399–410. [CrossRef]
2. Cervino, G.; Romeo, U.; Lauritano, F.; Bramanti, E.; Fiorillo, L.; D'Amico, C.; Milone, D.; Laino, L.; Campolongo, F.; Rapisarda, S.; et al. Fem and Von Mises Analysis of OSSTEM® Dental Implant Structural Components: Evaluation of Different Direction Dynamic Loads. *Open Dent. J.* **2018**, *12*, 219–229. [CrossRef] [PubMed]
3. Moraschini, V.; Poubel, L.; Ferreira, V.; Barboza Edos, S. Evaluation of survival and success rates of dental implants reported in longitudinal studies with a follow-up period of at least 10 years: A systematic review. *Int. J. Oral Maxillofac. Surg.* **2015**, *44*, 377–388. [CrossRef] [PubMed]
4. Ibáñez, J.C.; Tahhan, M.J.; Zamar, J.A. Performance of Double Acid-Etched Surface External Hex Titanium Implants in relation to One-and Two-Stage Surgical Procedures. *J. Periodontol.* **2003**, *74*, 1575–1581. [CrossRef] [PubMed]
5. Dawson, J.H.; Hyde, B.; Hurst, M.; Harris, B.T.; Lin, W.S. Polyetherketoneketone (PEKK), a framework material for complete fixed and removable dental prostheses: A clinical report. *J. Prosthet. Dent.* **2018**, *119*, 867–872. [CrossRef] [PubMed]
6. Najeeb, S.; Zafar, M.S.; Khurshid, Z.; Siddiqui, F. Applications of polyetheretherketone (PEEK) in oral implantology and prosthodontics. *J. Prosthodont. Res.* **2016**, *60*, 12–19. [CrossRef] [PubMed]
7. Thoma, S.D.; Cha, J.K.; Jung, U.W. Treatment concepts for the posterior maxilla and mandible: Short implants vs long implants in augmented bone. *J. Periodont. Implant Sci.* **2017**, *47*, 2–12. [CrossRef] [PubMed]
8. Raviv, E.; Turcotte, A.; Harel Raviv, M. Short dental implants in reduced alveolar bone height. *Quintessence Int.* **2010**, *41*, 505–509.
9. Felice, P.; Checchi, L.; Barausse, C.; Pistilli, R.; Sammartino, G.; Masi, I.; Ippolito, D.R.; Esposito, M. Posterior jaws rehabilitated with partial prostheses supported by 4.0 × 4.0 mm or by longer implants: One-year post-loading results from a multicenter randomized controlled trial. *Eur. J. Oral Implantol.* **2016**, *9*, 35–45.
10. Blanes, R.J.; Bernard, J.P.; Blanes, Z.M.; Belser, U.C. A 10-year prospective study of ITI dental implants placed in the posterior region. II: Influence of the crown-to-implant ratio and different prosthetic treatment modalities on crestal bone loss. *Clin. Oral Implants Res.* **2007**, *18*, 707–714. [CrossRef]

11. Urdaneta, R.A.; Rodriguez, S.; McNeil, D.C.; Weed, M.; Chuang, S.K. The effect of increased Crown-to-Implant Ratio on Single Tooth Locking Taper Implants. *Int. J. Oral Maxillofac. Implants* **2010**, *25*, 709–743.
12. Zurdo, J.; Romao, C.; Wennstrom, J.L. Survival and complication rates of implant-supported fixed partial dentures with cantilevers: A systematic review. *Clin. Oral Implants Res.* **2009**, *20*, 59–66. [CrossRef] [PubMed]
13. Salvi, G.E.; Brägger, U. Mechanical and technical risks in implant therapy. *Int. J. Oral Maxillofac. Implants* **2009**, *24*, 69–85. [PubMed]
14. Wolff, J.; Narra, N.; Antalainen, A.K.; Valášek, J.; Kaiser, J.; Sándor, G.K.; Marcián, P. Finite element analysis of bone loss around failing implants. *Mater. Des.* **2014**, *61*, 177–184. [CrossRef]
15. Cicciù, M.; Cervino, G.; Milone, D.; Risitano, G. FEM Investigation of the Stress Distribution over Mandibular Bone Due to Screwed Overdenture Positioned on Dental Implants. *Materials* **2018**, *11*, 1512.
16. Lauritano, F.; Runci, M.; Cervino, G.; Fiorillo, L.; Bramanti, E.; Cicciù, M. Three-dimensional evaluation of different prosthesis retention systems using finite element analysis and the Von Mises stress test. *Minerva Stomatol.* **2016**, *65*, 353–367. [PubMed]
17. Sahin, S.; Cehreli, M.C.; Yalcin, E. The influence of functional forces on the biomechanics of implant-supported prosthesis—A review. *J. Dent.* **2002**, *30*, 271–282. [CrossRef]
18. Mellal, A.; Wiskott, H.W.; Botsis, J.; Scherer, S.S.; Belser, U.C. Stimulating effect of implant loading on the surrounding bone. Comparison of three numerical models and validation by in vivo data. *Clin. Oral Implants Res.* **2004**, *15*, 239–248. [CrossRef] [PubMed]
19. Geng, J.P.; Tan, K.B.C.; Liu, G.R. Application of finite element analysis in implant dentistry: A review of the literature. *J. Prosthet. Dent.* **2001**, *85*, 585–598. [CrossRef]
20. Sohn, B.S.; Heo, S.J.; Koak, J.Y.; Kim, S.K.; Lee, S.Y. Strain of implants depending on occlusion types in mandibular implant-supported fixed prostheses. *J. Adv. Prosthodont.* **2011**, *3*, 1–9. [CrossRef] [PubMed]
21. Fernández Bodereau, E., Jr.; Fernández Bodereaum, E. Prótesis a Puente. Selección y valoración de places. In *Prótesis Fija e Implantes. Práctica Clínica*; Ed Avances Médico-Dentales SL: Madrid, Spain, 1996; pp. 295–320.
22. Fazi, G.; Tellini, S.; Vangi, D.; Branchi, R. Three-dimensional finite element analysis of different implant configuration for mandibular fixed prosthesis. *Int. J. Oral Maxillofac. Implants* **2011**, *26*, 752–759. [PubMed]
23. Turner, M.J. Stiffness and Deflection Analysis of Complex Structures. *J. Aeronaut. Sci.* **1956**, *23*, 805–823. [CrossRef]
24. Okumura, N.; Stegaroiu, R.; Kitamura, E.; Kurokawa, K.; Nomura, S. Influence of maxillary cortical bone thickness, implant design and implant diameter on stress around implants: A three-dimensional finite element analysis. *J. Prosthodont. Res.* **2010**, *54*, 133–142. [CrossRef] [PubMed]
25. De Almeida, E.O.; Rocha, E.P.; Freitas, A.C., Jr.; Freitas, M.M., Jr. Finite Element Stress Analysis of Edentulous Mandibles with Different Bone Types Supporting Multiple-Implant Superstructures. *Int. J. Oral Maxillofac. Implants* **2010**, *25*, 1108–1114. [PubMed]
26. Romeo, E.; Tomasi, C.; Finini, I.; Casentini, P.; Lops, D. Implant-supported fixed cantilever prosthesis in partially edentulous jaws: A cohort prospective study. *Clin. Oral Implants Res.* **2009**, *20*, 1278–1285. [CrossRef] [PubMed]
27. Céa, J. Approximation Variationnelle des Problèmes aux Limites. *Ann. Inst. Fourier.* **1964**, *14*, 345–444. [CrossRef]
28. Sotto Maior, B.S.; Senna, P.M.; da Silva, W.J.; Rocha, E.P.; Del Bel Cury, A.A. Influence of crown-to-implant ratio. Retention system, restorative material, and occlusal loading on stress concentrations in single short implants. *Int. J. Oral Maxillofac. Implants* **2012**, *27*, 13–18.
29. Bolle, C.; Felice, P.; Barausse, C.; Pistilli, V.; Trullenque-Eriksson, A.; Esposito, M. 4 mm long vs longer implants in augmented bone in posterior atrophic jaws: 1-Year post-loading results from a multicentre randomized controlled trial. *Eur. J. Oral Implantol.* **2018**, *11*, 31–47.
30. Toti, P.; Marchionni, S.; Menchini-Fabris, G.B.; Marconcini, S.; Covani, U.; Barone, A. Surgical techniques used in the rehabilitation of partially edentulous patients with atrophic posterior mandible: A systematic review and meta-analysis of randomized controlled clinical trials. *J. Craniomaxillofac. Surg.* **2017**, *45*, 1236–1245. [CrossRef]
31. Tabrizi, R.; Arabion, H.; Aliabadi, E.; Hasanzadeh, F. Does increasing the number of short implants reduce marginal bone loss in the posterior mandible? A prospective study. *Br. J. Oral Maxillofac. Surg.* **2016**, *54*, 731–735. [CrossRef]

32. Gastaldi, G.; Felice, P.; Pistilli, V.; Barausse, C.; Ippolito, D.R.; Esposito, M. Posterior atrophic jaws rehabilitated with prostheses supported by 5 × 5 mm implants with a nanostructured calcium-incorporated titanium surface o by longer implants in augmented bone. 3-Year results from a randomized controlled trial. *Eur. J. Oral Implantol.* **2018**, *11*, 49–61. [PubMed]
33. Chou, H.Y.; Müftü, S.; Bozkava, D. Combined effects of implants insertion depth and alveolar bone quality on peri-implant bone strain induced by a wide-diameter, short implant, and a narrow-diameter, long implant. *J. Prosthet. Dent.* **2010**, *104*, 293–300. [CrossRef]
34. Sahrmann, P.; Schoen, P.; Naenni, N.; Jung, R.; Attin, T.; Schmidlin, P.R. Peri-implant Bone Density around Implants of Different Length: A 3-year Follow-up of Randomized Clinical Trial. *J. Clin. Periodontol.* **2017**, *44*, 762–768. [CrossRef] [PubMed]
35. Peixoto, R.F.; Macedo, A.P.; Martinelli, J.; Faria, A.C.; Tiossi, R.; Ribeiro, R.F.; de Mattos, M.D. A Digital Image Correlation Analysis of Strain Generated by 3-Unit Implant-Supported Fixed Dental Prosthesis: An In Vitro Study. *Implant Dent.* **2017**, *26*, 567–573. [CrossRef] [PubMed]
36. Esfahrood, Z.R.; Ahmadi, L.; Karami, E.; Asghari, S. Short dental implants in the posterior maxilla: A review of the literature. *J. Korean Assoc. Oral Maxillofac. Surg.* **2017**, *43*, 70–76. [CrossRef] [PubMed]
37. Lemos, C.A.; Ferros-Alves, M.L.; Okamoto, R.; Mendonca, M.R.; Pellizzer, E.P. Short dental implants vs standard dental implants placed in the posterior jaw: A systematic review and meta-analysis. *J. Dent.* **2016**, *47*, 8–17. [CrossRef] [PubMed]
38. Block, J.; Matalon, S.; Tanase, G.; Ormianer, Z. Effect of Restorative Configurations and Occlusal Schemes on Strain Levels on Bone Surrounding Implants. *Implant Dent.* **2017**, *26*, 574–580. [CrossRef]
39. De Souza Batista, V.E.; Verri, F.R.; Almeida, D.A.; Santiago Junior, J.F.; Lemos, C.A.; Pellizzer, E.P. Finite element analysis of implant-supported prosthesis with pontic and cantilever in the posterior maxilla. *Comput. Methods Biomech. Biomed. Eng.* **2017**, *20*, 663–670. [CrossRef]
40. Ormianer, Z.; Matalon, S.; Block, J.; Kohen, J. Dental Implant Thread Design and the Consequences on Long-Term Marginal Bone Loss. *Implant Dent.* **2016**, *25*, 471–477. [CrossRef]
41. Villarinho, E.A.; Triches, D.F.; Alonso, F.R.; Mezzomo, L.A.M.; Teixeira, E.R.; Shinkai, R.S.A. Risk factors for single crowns supported by shorts (6-mm) implants in the posterior region: A prospective clinical and radiographic study. *Clin. Implant Dent. Relat. Res.* **2017**, *19*, 671–680. [CrossRef]

MDPI
St. Alban-Anlage 66
4052 Basel
Switzerland
Tel. +41 61 683 77 34
Fax +41 61 302 89 18
www.mdpi.com

Symmetry Editorial Office
E-mail: symmetry@mdpi.com
www.mdpi.com/journal/symmetry

www.ingramcontent.com/pod-product-compliance
Lightning Source LLC
LaVergne TN
LVHW070609170726
843515LV00004B/27